# THE RACE

# THE RACE

INSIDE THE INDY 500

## JAMES McGUANE

Sports Publishing books may be purchased in bulk at special discounts for sales promotion, corporate gifts, fund-raising, or educational purposes. Special editions can also be created to specifications. For details, contact the Special Sales Department, Sports Publishing, 307 West 36th Street, 11th Floor, New York, NY 10018 or sportspubbooks@skyhorsepublishing.com.

Sports Publishing® is a registered trademark of Skyhorse Publishing, Inc.®, a Delaware corporation.

Visit our website at www.sportspubbooks.com.

10 9 8 7 6 5 4 3 2 1

Library of Congress Cataloging-in-Publication Data is available on file.

Cover design by Tom Lau
Cover photo credit James McGuane

ISBN: 978-1-61321-915-7
Ebook ISBN: 978-1-61321-916-4

Printed in China

# DEDICATION

*To the memory of Tony Renna, Paul Dana, Dan Wheldon, and Justin Wilson—*

*four men who chose the racing life.*

# CONTENTS

# ACKNOWLEDGMENTS

SEEKING COMMENTS and stories from drivers who have won the Indy 500 was well down the list of what I had hoped to accomplish as I set out on this project. In time, I realized that what once may have been seen as a "sidebar" enhancement to a "photo essay" had, in fact, become central to the book.

I've always felt that a journalist or documentarian should avoid asking questions where the answer is already known. That notion seems to have served me well. I can think of no better way to learn of the racing world and the racing life. I greatly enjoyed the experience of "talking shop" with the entire (living) crop of past Indy 500 winners. There are qualities shared by each: honest opinion and recollection and a sense of humor. None of them "carefully measured their words." Each of them has my gratitude for their contribution to this book.

I was also well supported by an elite assortment of individuals—the "pros" of racing. Many are former Indy drivers who continue in the sport in other capacities.

Stefan Johansson, Pancho Carter, Roger Penske, Lyn St. James, Larry Foyt, Sam Schmidt, Lee Kunzman, Chip Ganassi, Bill Simpson, Sarah Fisher, Michael Andretti, and Scott Goodyear are particular standouts.

Race teams are particularly busy in the pressure-filled month of May around Indy. The teams' decision to graciously make time for

me—notebook and camera at hand—was invaluable to me. Thanks is extended to Mo Nunn Racing, Andretti Autosports, Panther Racing, Team Penske, Schmidt-Peterson Motorsports, Ganassi Racing, Dreyer & Reinbold Racing, A.J. Foyt Racing, Rahal/Letterman/Lanigan, Newman/Haas, Vision Racing, CFH Racing, Hemelgarn Racing, Cheever Racing, KV Racing (and a lot more).

I was very happy to discover in 2003 that my Notre Dame classmate Mike Griffin was a partner in Panther Racing—a top echelon, championship team. Mike has opened many doors, answered a thousand questions with wit and wisdom, and been a durable friend. Thanks, Mike!

I have to recall just how slight my knowledge of racing was as this project began. I was neither a "motor head" nor a "gear head." I was simply a fan—early on, listening to the annual race on the radio, hardly believing that one could drive five hundred miles in a single day. My uncle Jim was a friend of Jim Rathmann and he secured two infield tickets—one for me and one for my brother Frank—for the 1965 race. We got to see Jim Clark win, the first victory in a rear engine car. It was also Mario Andretti's rookie year.

There have been hundreds of people, events, and situations that have given me the courage to call this an "inside" look at Indy racing. In no particular order, here are some that come to mind and have my thanks:

David Storvick—IndyCar Ministry; Uniden/Bearcat Scanner; Hoosier 100; Indy Hostel; Ron Hussey—Houghton Mifflin Harcourt Company; Brigid Kavanagh for a gift to Skip Barber Racing/Lime Rock Park; Dick Wolfe; Phoenix International Raceway; Bombardier Racing; Nazareth Speedway; Kika Concheso; Ann Fornoro; Susan Bradshaw; Jane Barnes; Bob Linda; Joe Crowley; Jeff Dennison; Aaron Nelson; Kelby Krauss; Jeffrey Rupple; Hasselblad Camera Company; Ziggy Harcus; Phase One/Capture One/Digital Transitions/Jeff Lin and Allison Watson; Zeiss Lenses; Indiana Oxygen; Lincoln Welding; Holmatro Safety Team—Mike Tate; Delphi; Patty Reid; Dale Ratermann; Chuck Brewer; Jeff Troyer.

Chicagoland Speedway, Texas Motor Speedway/Eddie Gossage; Joie Chitwood; Randy Bernard; Worldwide Pants; ABC/ESPN; Rodger Ward Jr.; Indianapolis Star; Mark Ferner; Brian Figg; Derek Bell; Danica Patrick; Xtrac/Andrew Heard; Panoz G-Force /Simon Marshall; Dallara/Sam Garrett; Firestone/Jimmy Russell and Dale Harrigle; Toyota Racing Development; Chevy Racing/Chris Berube and Judy Kouba Dominick; Sunoco; and Verizon.

I'd like to single out Honda Motorsports and Honda Performance Development for the particular help and encouragement that they've provided me over the entire course of this book project. I found them to be superb engineers, nice people, and sharp racing competitors.

T. E. McHale of Honda Motorsports has my very deep gratitude.

Parachutist on Race Morning

Thanks also to the amazing Hoosier hospitality that the Speedway people have provided over the years: Deb Taylor, Buddy McAtee, Ron McQueeney, Donald Davidson, Dr. Henry Bock, Dr. Timothy Pohlman, Tom Carnegie, Dan Edwards, Bill Walters, Tim Sullivan, Dave Brown, Jeff Horton, Kevin "Rocket" Blanch, Derrick Walker, Brian Barnhart, and the Hulman and George Families.

My editor, Julie Ganz at Skyhorse Publishing, has been a bright and patient leader as we sorted through 7,000 of my film and digital images and managed to assemble interviews with every living Indy 500 winner. I have been guided through the publishing schedule by the sure hand of Editorial Director Jay Cassell. Thank you, Julie and Jay!

Finally, I'm grateful to Andrew Stuart and Paul Starobin at the Stuart Literary Agency for their encouragement and hard work in helping me to shape this documentary project.

It's been said of journalists that they "learn in public." I'm grateful to the many true pros in the racing world who have inspired and encouraged me: Jack Arute, Allen Bestwick, Curt Cavin, Eddie Cheever, Chris Economaki, Scott Goodyear, Don Kay, Michael Knight, Jamie Little, Larry McReynolds, Robin Miller, Paul Pfanner, Dr. Jerry Punch, and Darrell Waltrip.

Friends and family have supported me in ways that they may never know. Thanks to Friends and Family: Sheila, Katie, Meg, Jamie, Brooke, Tom A, Karl, Mano, Judy, Clare, Don A, J O'B, Ritchie A, Kevin, Robby, Mary M (Apt 5C), George Griffin, Dede B, Joe Iron, Trudy, Frank, Patty and Joey, Paul Landis, Roy Beauchamp, Paul Goldsmith, Stephen Ives, Roma Agrawal, Shawn Dougal, Eric at Yellow Umbrella Books, Clayton Calderwood, Bill Marcato, Min-o, Pamela Rosen, Joe'l, Jon Harrison, Bob Wagner, Robert Parker, Rick M, Joanne T, Massamba, Bob Staats, and Jose Martos Gallery.

My initial contact at the Speedway, Fred Nation, has been a stalwart supporter. His insight, inspiration, patience, and friendship greatly helped to get this book between two covers. Thank you, Fred!

The Borg-Warner Trophy.

# INTRODUCTION

**THE INDIANAPOLIS** Motor Speedway has been the premier venue for auto testing and racing for one hundred years. Its longevity has been assured by man's fascination with both technology and speed.

In 1909 the term "horseless carriage" had a derisive ring to it. As the automotive age got "up to speed" (so to speak) the concept of "horsepower" was no longer a laughing matter.

The two-and-a-half-mile track at Indianapolis was born of a need for tinkerers, inventors, and other pioneering auto "manufacturers" to "torture test" their creations. The first official 500-mile race took place in 1911. There was little consistency among the cars that entered that race, save the fact that each had four wheels and an internal combustion engine. It was a noisy, smoky, oily, and deadly spectacle. The majority of the forty cars that started the race were from the United States, but the event billed as The International 500 Mile Sweepstakes Race truly was international. There were cars and drivers from France, Italy, the UK, and Switzerland. The family name of the Swiss driver would go on to be famous around the world: he was Arthur Chevrolet. A look at the names of the car models and their manufacturers that entered that first race indicates where the nascent automobile industry was coming from and where it would be headed. The entry card bears names such as Mercedes, Stutz, Fiat, Benz, and Firestone-Columbus (the

entrant's official name was the Columbus Buggy Company). Several cars bore the name "Case." This was a rare foray into automobile manufacturing for the company that began in 1844 as the J. I. Case Threshing Machine Company. Case would go on to become a giant in the production of tractors and other agriculture and construction equipment. The engine that they raced at Indy was an experimental design that burned "oil."

The first "500" set the standard as *the* race to showcase the best driver in the best car, with "up to the minute" technology. The Speedway to this day proves to be a "laboratory" for automotive and racing innovation. You could say the 1911 race was won as a result of a breakthrough innovation. In those early years it was customary to race with *two* men in each car—the driver and a "riding mechanic." On some cars this mechanic would operate a lever to pump lubricating oil through the engine (automatic oil pumps were not yet in general use). He would also function as an on-board "spotter" to keep the driver aware of surrounding traffic.

Ray Haroun was entered in a US-made Marmon "Wasp" race car. In the interest of saving weight he was alone in the cockpit. His team had rigged what would become an automotive standard—a rearview mirror. It was the first such recorded use. Haroun would win the race—and the winner's purse of $10,000.

The initial 500-mile race in 1911 also saw the first of many Indy fatalities. About twenty-five miles into the race, the car driven by Arthur Greiner lost a front wheel and crashed during the second turn. The driver and riding mechanic were thrown from the car. Initial reports said that riding mechanic Sam Dickson was "killed instantly," a fact that has come into question. The crowd swarmed around both Greiner and Dickson. The state militiamen who were charged with security at the racetrack had to use their guns as clubs in order to clear a path for the medical attendants. In the "show must go on" tradition, the track debris was cleared and the race was resumed.

The Marmon Wasp, the winning racer at the first Indianapolis 500.

In looking at a century of racing at Indianapolis, there are *decades* where nothing changes very much. Then, a single innovation—the rear engine, the aerodynamic "wing," or innovative tire composition—can revolutionize the sport. It causes the teams to scrap the equipment they *were* using and experiment with the *new*. A team with sponsors with "deep pockets" can become quite innovative. Some little trick that saves a half second on each lap—well, there are 200 laps in the 500-mile race.

Over a span of twelve years I visited the Speedway eight times for the race weekend—a Memorial Day tradition. Veterans, past and present, are honored each year. There is always some grand "showing of the flag" on the apron of "pit lane" prior to the race. Indy knows how to make those in the armed forces feel appreciated.

The venue is *huge*. The racetrack is a "squared oval," two-and-a half miles around. The "infield" is an anthill of distinct neighborhoods: first and foremost is Gasoline Alley. This is where qualifying teams garage, service, repair, and show off the cars they are preparing and are tweaking into "race trim." The fan (with the right access pass) gets a *very* close-up look at his or her heroes and their machines. There are "perks" for the top-ranked teams. The garage's proximity to the track entrance is based on a "points" system. Aside from "location," there is a neat consistency to all the garages. The teams roll in *all* of the tools and parts that they need for racing. When they leave, it's a blank slate for the next race.

Each of my early visits to the Indy 500 would end with the knowledge that more "coverage" would be needed before weaving it into the desired book. After four or five additional sessions at additional 500s, *change* was complicating matters.

Initially a team would go with Chevy, Toyota, or Honda for an "engine package." They also made a choice for "chassis"—either Panoz-G Force or Dallara. This is basically the whole race car, less the aforementioned engine. Everyone has Firestone tires. Everyone has an Xtrac transmission. Then the "automobile crisis" happened!

The marching band will return shortly.

The photographs here were collected between 2003 and 2015. I began with the notion that covering the great Indy 500 race for a few years would yield material that was worthy of publication. Like all true documentaries, once begun, the artist finds himself merely "along for the ride."

Early enthusiasm moved me to follow the teams that had competed in the 500 to a few other venues in the Indy Racing League season. I made trips to Chicagoland in Illinois, California Speedway in Fontana (now know as Auto Club Speedway), Phoenix International Raceway in Arizona, Texas Motor Speedway in Ft. Worth, and Nazareth Speedway in Pennsylvania. These subsequent races provided me with valuable access to the cars, the teams, and the racing life. For five consecutive years I returned to Indy. Following each of those trips I reshaped my portfolio and explored opportunities for publication. Personal and professional projects intervened and left some gaps in my coverage. For the record, the years covered are 2003, 2004, 2005, 2006, 2007, 2009, 2014, and 2015.

Chevy stopped racing (GM was in big trouble). Toyota didn't come to America just to compete against *another* Japanese engine, so they dropped out.

Honda, heroically, "stayed the course." They were the supplier of engines to every team. A Honda executive quipped, "Yes, we're sure to win the Indianapolis 500, but we also know we'll come in *last*."

Drivers changed teams. Drivers retired. Drivers died. Drivers tried NASCAR. Drivers lost their "ride." Sponsors, peripheral equipment, and rules all changed.

The severe pangs in the world economy put pressures of change on the Speedway and Indy racing in general. Gradually it became clear that this was not a tight linear story; this is *about* change.

Finally, in early May of 2015, the project became a reality. The pictures shot at the ninety-ninth running of the Indianapolis 500, when added to all the previous images, brought the total to over seven thousand. The author appreciates the full spectrum of (living) Indy winners who gave time and effort to this project. A portrait of Indy emerges through their words and thoughts that no writer could fashion.

The changes, confusion, conflict, and chaos surrounding open-wheel racing during my purview has not diminished my fascination with this great American competition.

\*\*\*

This gentleman came to the racetrack by jumping from a perfectly good aircraft.

## CHRIS ECONOMAKI
### OCTOBER 15, 1920–SEPTEMBER 28, 2012

*The place of honor (first interview) is held by Chris Economaki, may he rest peacefully. Chris's insight on both racing and journalism proved invaluable to this project.*

*Chris was known as the "Dean of American Motorsports." In 2006 he saw his name affixed to the Economaki Press Conference Room at Indy. When he sat for an interview in 2004 he reflected on racing, his career, speed, and Indy.*

Television makes you "familiar" to a lot of people—and I was on television for thirty-four years, longer than any other sportscaster. Then, if you're a hard-core racing fan you read my newspaper—every deep-down race fan reads *National Speed Sport News*—which I brought up to prominence. It's the publication that covers the full breadth of automobile racing, not only in this country but also around the world. So that's how people know me.

Before the term "crew chief" was created, I was crew chief on a midget racing car throughout the 1938 and 1939 seasons. I sold newspapers and wrote race reports at the same time. The term "journalism" hadn't been created yet.

### ON EARLY RACING IN NEW YORK CITY:
Well the first race in New York City was in Van Cortlandt Park (Bronx).

After that—the chronology of events went like this:

The *Madison Square Garden Bowl*, which was across the street from LaGuardia Airport, raced in 1936. Two armories in Manhattan had races. You had Kingsbridge Armory in the Bronx, at 181st Street or something.

The original Madison Square Garden had races. There were races in Brooklyn, at Atlantic and Bedford avenues, at an armory there. There was the Bay Ridge Oval out in Brooklyn. Up near the Queens line there was the Farmers Oval. The Coney Island Velodrome. The Sheepshead Bay Board Speedway . . .

### WHAT ELSE?
Well, there were a whole lot of races.

In fact, in the early 1900s some guy wanted to see how fast his car would go, so he got the American Automobile Association to set up a "measured mile" on the first paved street in Brooklyn—Coney Island Boulevard—and he drove down the street at the incredible speed of sixty-five-and-a-fraction miles per hour (I think it was). Unbeknownst to him—or anybody else—it became the world land speed record.

In Brooklyn!
Yeah!

Arthur L. Carter Sr.—the last Hoosier who served in World War II with the Tuskegee Airmen at the 2009 race. Sadly, Arthur passed away in January of 2015 at age 92. The following facts gleaned from the pages of the *Indianapolis Recorder* newspaper (where Carter worked in later years) speak for themselves:

*"Between 1941 and 1946 more than 1,000 Black pilots were trained at Tuskegee, Ala. Collectively, the Tuskegee Airmen received Presidential Unit Citations, 50 Distinguished Flying Crosses and eight Purple Hearts.*

*"Thirty of the pilots were from Indiana and 13 of them, including Carter, were graduates of Crispus Attucks High School. Sixty-eight Tuskegee Airmen were killed during the war; 32 were captured as prisoners of war."*

This flyover by two vintage B-25 Mitchell bombers, named *Special Delivery* and *Take Off Time*, is in tribute to the Tuskegee Airmen.

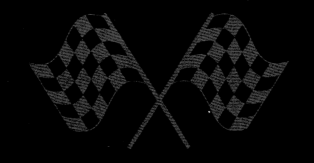

# PART I

# THE RACE COMPONENTS

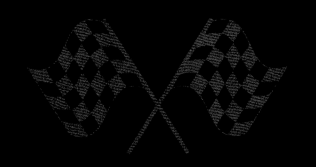

# 1
# THE RACE CAR

WHEN THIS project began in 2003 there were two approved chassis manufacturers—Panoz G-Force and Dallara. A race team cast their lot with one or the other. A multiyear development for a newer and safer vehicle saw Dallara emerge as the sole supplier of the Indy chassis starting with the 2013 season. The approved "package" that they offer includes optional equipment, aerodynamic variables, and spare parts. Over the one hundred years of Indy racing the car has gotten far lighter, lower, and sleeker. Even the driver's helmet and faceplate are now shaped to offer as little air resistance as possible. Comfort is but a minor consideration. The car is meant to knife through the air with just enough "downforce" to keep the powerful rear wheels in contact with the asphalt and provide the front wheels with the precise amount of "bite" for high-speed steering and handling. Vast effort has been expended in pursuit of the racer's "holy grail"—a perfectly balanced race car.

The car is already in place on the starting grid but the team has removed one of the sidepod panels to work on the car.

Carbon fiber *will* conduct electricity but not very efficiently. The metallic foil seen on the underside of the panel is to provide a good electrical "ground" for the radios, transponders, and other electronic gear.

The nose and tail sections are seen mounted on a stand behind the pit wall. They stand in reserve. Each provide a "one piece" renewal of a damaged part. Teams endlessly practice such substitution. Dallara estimates that if the front wing assembly, which is made of carbon fiber, were made of steel, it would be more than three times the weight. It would reduce "lap speed" at Indy by 1.5 mph.

The race cars, in the interest of reducing weight, leave the battery and starter motor back at the pit. At the beginning of the race or in the event that the car stalls during a stop, the starter is right at hand. The long probe is geared at the very end and meshes with gears in the transmission. Danny Sullivan, in a separate chapter, recounts a fortunate occasion where he is able to "jump start" his stalled (and speeding) race car.

**Fuel Bladder.** This unit is not fabricated by Dallara, but comes from a separate supplier. The technology is similar to that used in helicopters. Dallara builds a whole range of race cars and what is shown may or may not be "IndyCar" specific. Certain configurations feature a "buckeye" fill port on each side of the car for use on tracks that pit to different sides. Fuel capacity in IndyCars during the span of years covered by this project have ranged from 30 gallons of methanol down to the current capacity of 18.5 gallons of E85. "Baffles" in the tank control the sloshing brought on by the g-forces. The "pickup" strives to make every drop of fuel in the tank available to the racer. Drivers are constantly reminded to "reset the fuel indicator" at the end of a pit stop for more accurate "end of race" and "scheduled stop" strategy.

Here is the "buckeye" on a 2015 car. It's obviously on the left side, closest to the pit wall. Safety is designed into the entire fuel system in several ways. The port snaps shut and locks as the fill probe is removed. In the event of a wreck, the supply line from the tank to the engine stops pumping.

Jacob of A. J. Foyt Racing is using a mini-grinder to shave down a chassis component so that it joins its adjacent part with almost no gap. The tiniest mismatch between parts can affect aerodynamic efficiency. If the part is on the topside of the car where it is visible it will have its vinyl "skin" attached.

The car is constantly assembled and reassembled for repair, maintenance, and a dozen other reasons. Here's a look at the "underside." The driver's cockpit has its own floor, and that's not what is shown here. Air is directed under the car into what are known as "tunnels." Part of the race strategy involves "tuning" the airflow in that area under the speeding car for maximum effect. The "ground effect" bargain is well known. Get it right, and the car will "stick" to the course in the corners. Overcorrect in that direction, and the "drag" will slow the car on the straights.

The nose assembly here demonstrates the idea that IndyCar racing is more like NASA than NASCAR.

The "livery" on Oriel Servia's #32 Rahal-Letterman-Lanigan car has a special #THANKSDAVE message honoring the *Late Night* host's retirement from TV.

This is a rare opportunity to see the axle, wheel hub, brake rotor, wishbones, and other parts of the suspension. The wishbones are steel alloy covered with a skin of carbon fiber and resin.

A. J. Foyt Racing has removed the entire rear section of the car to get at the innards.

Here the mechanic from the Schmidt-Peterson team is holding the ultra-lightweight carbon-fiber brake rotor covers. They take a good bit of abuse as tires are changed. Small ends can crack or break off and a constant chore is to epoxy them back to perfection. Each weighs in the neighborhood of eight ounces.

The seatback (as well as the steering wheel) is removable to allow the driver easier access to the cockpit. He (or she) will ease down into the custom molded seat; the helmet and "head and neck" protector (one brand is called the HANS Device) would already be in place. This piece becomes the "coaming" of the race car. There is padding both behind the driver's helmet and also along the right side. When racing on ovals the powerful centrifical force pushes the driver mainly to the right.

The "T" car. Here the letter next to the #4 identifies the car as Panther Racing's "backup" racer. It was identical to the main entrant in every way. Rules permitting, it could be put into service to replace a damaged car. There is a quilted black drape shielding the day's "aero setup" from the prying eyes of rival teams. IRL rules have now outlawed such "black magic" gamesmanship, in the years after this photo was taken.

The unit being installed above the air intake port is the housing that contains the "car cam." The camera transmits the video (and sometimes audio) information, in real time, to both a dedicated recorder as well as to the broadcasting team. Its point of view is usually the same as that of the driver, but some cameras can be diverted to a rear view.

When the technology was first introduced, live cameras would be installed only on the cars of a few top drivers, presumably because they were more likely to be in the race until the finish. Cars not so selected would be mounted with a "dummy cam" so that everyone shared the extra aerodynamic "drag" that the unit would cause.

The car cam had an ingenious method of self-cleaning the lens. The aperture, which was no larger than a pea, could be totally obliterated by a single drop of moisture, oil, insect, or (most likely) a flying, sticky blob of tire rubber. Thus, a spool of clear tape was threaded in such a way that it became the "leading edge." As the broadcast production crew saw fit, the tape could be advanced an inch or so, moving the obstruction away and clearing the view.

Interior of camera unit without housing.

This savvy "over the wall" tire changer keeps a spare wheel nut loosely clipped to his belt. If one is dropped or gets away during a pit stop he's ready.

**Paoli wrench:** This air-driven workhorse (made in Italy) does one thing very well: It engages with the wheel nut to mount or dismount a wheel and tire. It may *seem* as if it has been carelessly "cast aside" as the tire changer acts out his meticulous ballet, but the designer knew the abuse the tool was likely to see. The precision of the front "teeth" is shielded by a sturdy steel "collar." An experienced hand can send it safely skidding across the pavement when the task has been completed.

Looking down at the brake rotor, showing wheel hub that receives the nut.

**Signboard:** Reset Fuel #21. There is always a chance that modern telecommunication methods can fail or malfunction. Here is an "in your face" reminder. Accounting for every drop of fuel is essential. Races are won or lost depending on how well fuel use is calculated.

**Buckeye on white Toyota.** The Buckeye is designed to mate perfectly with the fuel probe before a drop of the gravity-fed fuel is allowed to pass. Obviously, the driver is inches away from the operation. A fireman is closely engaged with the race team during each pit stop. He holds a canister of pressurized liquid, which he can direct to even the *suspicion* of a small spill. (It can be difficult to see an alcohol flame.) In the event of an actual fire, the crew is well trained in fire suppression. Extinguishers are at hand just over the pit wall. A two-inch fire hose is fully pressurized and ready; one hose serves two adjacent pits.

Transporters, of necessity, use every bit of storage space as they truck from race to race. Here, the lift gate retrieves the car from one of the storage bays in the "attic."

This is a rear brake duct backing plate for the 2003 generation IndyCar from Panoz G-Force (other side shown below). It's extremely light and quite rigid. The reverse side shows embedded metal "taps" for connecting screws or bolts. Each has a bit of "wiggle room" for a snug, secure fit.

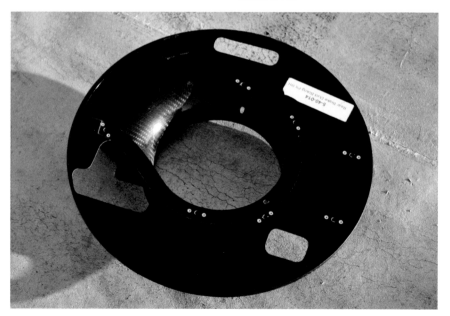

# Q&A WITH SIMON MARSHALL
## CHIEF DESIGNER, PANOZ G-FORCE

*Marshall spoke with me in October 2003 and again in February 2004.*

### Q: CAR CHANGES FOR NEXT YEAR (2004)?

They've taken away some downforce from us in a very efficient area, which is the "underbody." They've reduced the diffuser height in the back of the car, which is the driving force behind sucking the air into the under-wing area.

### Q: EFFICIENT? BECAUSE YOU WISH THEY WOULDN'T TAKE THAT AWAY FROM YOU?

Yes, it gives a lot of downforce with a small amount of drag, whereas other features on the car—even the wings—are not that efficient.

### Q: WHY WOULD ONE SPECULATE THAT THEY (THE IRL) WOULD DO THAT?

Oh, because they know that the cars always seem to be going faster and faster, and I think—you see they regulate the speed at the tracks with different wing angles and different "gurney" or "wicker" heights—and they can take five miles per hour off the cars if they feel we're going a bit too quick. You've gotta have Indy as being the fastest track just for P/R reasons, and um, the insurers do not like them going over 230 miles per hour.

### Q: INSURANCE?

Well yes, someone has to insure all these people sitting in the stands. These things have gone out—into the stands before. Someone has got to pay someone an awful lot of money.

Money. It just comes down to *cash* at the end of the day. So they've taken some efficient downforce on the car, knowing that for us to put it back, we have to look in *different* areas to put it back.

### Q: THEY KNOW YOU'LL WORK AROUND THIS?

They know we will. It's just, every year, they have to "take away" from us—and we try to get it back.

### Q: WE'LL SEE A WHOLE NEW "SHELL" OF A CAR NEXT YEAR?

The rear wings'll be the same, because they're mandated. Everything else will change.

### Q: CAN YOU POINT OUT THE DIFFERENCES BETWEEN THE IRL AND FORMULA ONE?

In essence, a hell of a lot more money. That's all it would take, is more money. They (Toyota) are putting their car together in Cologne at the moment. They are currently campaigning that car in F-1. It takes a thousand people to put that car on the track. That's for two drivers, and they're up against Ferrari and Williams. They've all got in excess of $100,000,000 to spend on the car.

### Q: HERE IN THE IRL THERE ARE ABOUT A HUNDRED PEOPLE BEHIND EVERY DRIVER.

Well there's a *few* of us doing the designing and building of it [*laughs*]. The teams have, uh . . . well I suppose Ganassi's got tons of those people, a hundred, I'd say.

### Q: REALLY, ONLY TWO CHASSIS MANUFACTURERS?

The third one that's been approved is Falcon Cars, which was a hurriedly put- together outfit. They swung a deal with the IRL to be the third manufacturer. The teams aren't stupid—they could see what was a good bet, and what isn't. And no one signed up for their car; therefore they went away, even before they even produced a car. This series can't support any more than two—because of markups on the parts.

### Q: CAN YOU DESCRIBE VISUAL DIFFERENCES BETWEEN G-FORCE AND DALLARA, BESIDES THE SHAPE OF THE AIR INTAKE PORTS?

We've both got lumps over the front dampers. They've got a bulbous piece in front of the windshield, and a lower nose, and a "pull-rod" system—a pull rod on the front suspension. Of course we've got a push rod on our suspension.

## Q: TEAM PENSKE IS USING BOTH CHASSIS. DOES ANY DIFFERENCE SHOW UP IN THEIR PERFORMANCE?

I think, at the moment, that the two cars are *so* close that it's hard to actually call between them. Which has ended up in this series being good for both us and Dallara this year. Also the engines are fairly close, although that has changed and "swung around" during the season. Recently it's swung around into Chevy's favor because they've just got a new engine now, which they got "passed" somehow. It was approved, somehow. It wasn't meant to be, but they were looking so bad before—it's not in the IRL's interest to make Chevy look bad. Chevy puts a lot of money into the series, as do Toyota and Honda—they need to get lots of positive press out of it. At the end of the year, everyone can do some pretty good ads saying how great they are.

## Q: WHAT CAN YOU SAY ABOUT THIS TEXAS TRACK (TEXAS MOTOR SPEEDWAY)?

You can't really pull ahead easily on this sort of track and with the cars in this configuration. The guys "behind" find it easier to, kind of, "pull alongside," but once they *get* alongside they can't get past because of the increased drag. You fall back behind in the "slipstream." So the cars are going to rush around here "slipstreaming" each other, but in a pack, which will make it exciting. A bit dangerous, but . . .

**[Kenny Brack was badly hurt that day, a day marked by several wrecks.]**

## Q: IS THIS TEXAS TRACK UNIQUE IN ITS AERO QUALITIES?

It's fairly unique. It's got such high banking. The teams can take off a lot of downforce and drag here. You don't want the drag. If you take off too much downforce, to try to lose drag, you end up sliding, slightly in the corner, and reducing speed in the corner. And then your "lap time" doesn't improve. It's faster down the straights and slower in the corners.

# 2
# THE ENGINE

**WHEN MY** research began, Chevy, Honda, and Toyota were the three engine suppliers for Indy racing. Worldwide economic woes brought on a period where only Honda was supplying engines. Now, happily for true-blue American race fans, General Motors' Chevy is back in competition.

For decades, hopeful racers who had teamed up with a mechanic or two would tinker with cylinders, rods, heads, valves, plugs, crankshafts, and a thousand other engine parts, pack their innovations under the hood of their race car, and set off for "qualifications trials" at Indy. They faced a simple truth: the grueling 500-mile contest will *only* reward power (performance) when it's coupled with reliability. This innovation and tinkering is now carried out by billion-dollar enterprises like General Motors (Chevrolet) and Honda. The parts that make up the modern Indy racing engines are masterpieces of metallurgy. The pistons are "billet machined aluminum alloy." The crankshaft, bearings, and connecting rods are "machined alloy steel." The engine is described as a 2.2-liter twin-turbocharged V6. Teams can make modifications that include "electronic throttle control," fuel injection, and other aspects of "electronic engine control" but are forbidden by the manufacturers from meddling with the internal engine parts. New engines are assembled at the factory and are delivered to the teams "ready to race."

When the engine approaches its 600-mile (or so) "racing life" it is packed in its original shipping case and returned to the manufacturer. The team's contract with the manufacturer allows for practice, testing (time trials), qualifying, and actual races. Spare engines can be standing by to replace those damaged in crashes.

A cutaway model of Chevy's earlier V8 IRL engine.

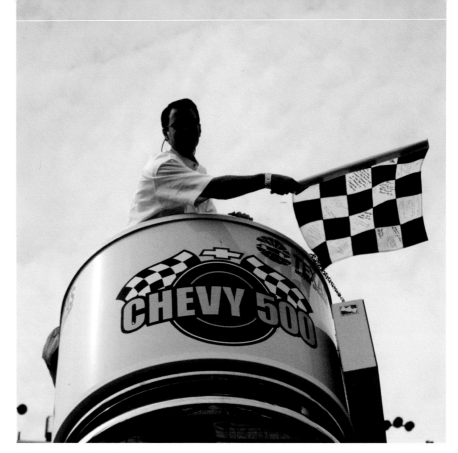

Chevy was the "naming sponsor" for the 2003 500-mile race at Texas Motor Speedway.

The engineers from Chevy Racing are working "shoulder to shoulder" with Team Menard.

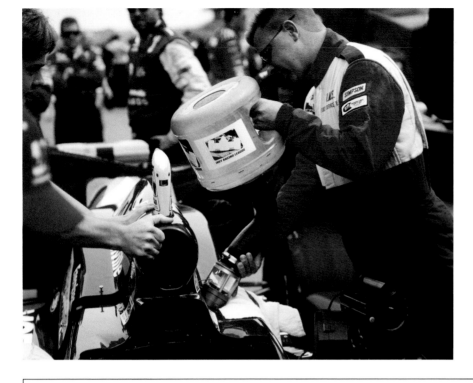

For qualifying the teams wish the car to be as light as possible and therefore tow a virtually empty race car to the track, where the IRL officials dispense the minimum required for the attempt—currently four gallons of ethanol.

| 2003 CHEVY INDY V-8 SPECIFICATIONS | | 2003 CORVETTE Z06 ENGINE | |
|---|---|---|---|
| **Specifications** | | **Specifications** | |
| Displacement: | 3.5 liters (214 ci) | Displacement: | 5.7 liters (350 ci) |
| Horsepower: | 675 hp @ 10,300 rpm | Horsepower: | 405 hp @ 6,000 rpm |
| Fuel | Methanol | Fuel: | 93 octane unleaded gas |
| Redline: | 10,300 rpm (per IRL rules) | Redline: | 6,500 rpm |
| Compression Ratio: | 15:1 | Compression Ratio: | 10.5:1 |
| Bore Diameter: | 93 mm (3.66") | Bore Diameter: | 99 mm (3.90") |
| Crankshaft Stroke: | 64.4 mm (2.53") | Crankshaft Stroke: | 92 mm (3.62") |
| "V" Angle: | 90 degrees | "V" Angle: | 90 degrees |
| Valvetrain: | Dual overhead cams | Valvetrain: | Pushrods with overhead valves |
| Valves per Cylinder: | Four | Valves per Cylinder: | Two |
| Camshaft Drive: | Gears | Camshaft Drive: | Chain |
| Cylinder Case Material: | Aluminum | Cylinder Case Material: | Aluminum |
| Cylinder Liners: | Wet light alloy | Cylinder Liners: | "Dry" iron |
| Cylinder Head Material: | Aluminum | Cylinder Head Material: | Aluminum |
| Crankshaft Design: | 180 degrees | Crankshaft Design: | 90 degrees |
| Fuel System: | Sequential EFI | Fuel System: | Sequential EFI |
| Throttle System: | Individual runner | Throttle System: | Throttle body |
| Lubrication System: | Dry sump | Lubrication System: | Wet sump |

These are Chevrolet engine specifications comparing the 2003 Chevy IndyCar V8 engine with the 2003 Corvette Z06. At that time, Honda's and Toyota's engines would similarly conform to the IRL's design parameters.

*How do motor oils and gear oils function in racing engines and transmissions? How are they put together? What additives are in there and in what concentration? What is the difference between synthetic and mineral-based oil?*

*Early in this project I was fortunate to connect with Mark Ferner, an engineer with Shell. In a series of conversations, Mark very succinctly describes the drama taking place deep inside a speeding race car. Mark is refreshingly candid. At the same time he discreetly maintains the confidences he enjoys with various race teams, designers, and with the Speedway.*

Some race teams may consider oil a "non-variable." You put it in, you know it works, end of story. Perhaps they think it's one less thing to worry about. However, there are some subtleties with oil formulations that can help you get more of the engine's power. This is true in both qualifying and actual race conditions.

In years past, many race teams such as those in the [Indy Racing League] or NASCAR would start out with an SAE 20W-50 conventional, mineral-based motor oil. This is one of the thicker viscosity grades. A thicker lubricant is generally thought to be safer and more conservative than a thinner viscosity lubricant. So SAE 20W-50 was a good starting point for many race teams. It's readily available, not that expensive, and teams have likely used it as they worked their way up to the higher levels of sponsorship.

A straight weight SAE 50 is generally considered to be a thick oil at engine operating temperatures. It will thicken as it cools down to room temperatures. It will further thicken at subzero temperatures. However, if an oil is *too* thick to pump through the engine, there is a risk of oil starvation. Here, precious engine power would be needed to force thick oil through small passages. Likewise, it will thin when heated—much like when cold maple syrup is difficult to pour when cold but thins quickly when placed in the microwave oven. Motor oils also thin when heated, but multigrade oils will thin less when heated. To that point, an SAE 20W-50 oil will look like an SAE 20W when cold

but will thin in viscosity when heated to appear/behave more like an SAE 50 oil at operating temperatures. With a more consistent overall viscosity, it's easier for an engine builder to design the hardware tolerances/clearances and set oil pump specifications.

Most lubricating oil formulations are a carefully balanced recipe of base oils and a variety of additives. The base oils typically make up most of this formulation. The additives could include viscosity index improvers, antiwear additives, antifoam additives, detergents, dispersants, and friction modifiers. It's imperative that the various base oils and additives should *complement* each other—not *contradict* or interfere with each other.

## Mineral versus Synthetic Nomenclature

The selection of the base oils used to be pretty cut-and-dried. You'd begin with crude oil, put it through a refinery, and the resulting products were "conventional, mineral-based" base oils. Then came the man-made molecules, which were historically the polyalphaolefins that most people referred to as "synthetic" base oils, but some esters were also referred to as synthetic base oils. In recent times the refining process has been enhanced so much that even some *mineral* products from crude oil that are run through certain refining processes long enough and hard enough may end up having the qualities of a *synthetic* oil. For that reason there are many oils on the market today that say "synthetic" on the label and have originated from crude oil.

One key lubrication area in any engine is where the camshaft meets the follower. The camshaft has a series of "egg-shaped" lobes. As the lobes rotate, they push the follower against the spring-loaded valves at the right times to allow the engine to breathe. The engine breathes in air and fuel and then exhales exhaust gasses. The points of contact where the lobes push open the spring-loaded valves is only a small line of physical contact and springs keep the valves closed normally. In racing situations with such high rpm, a lot of times they will use very high spring pressures to keep the valves closed. The race teams don't want the springs to be too low in pressure, lest the valves "float" in an open position longer than desired at high engine speeds. If you do the math, the spring pressure acting on this small line of contact would calculate to about 200,000 pounds per square inch! Two hundred thousand PSI is enough to squeeze away the base oil, be it synthetic or mineral or a synthetic blend of the two.

A very common type of antiwear additive found in motor oil formulations is a zinc compound with phosphorus and sulfur and is spelled with several syllables (a common molecule is zinc dialkyldithiophosphate). Hence, many people just refer to it in the shorthand "zinc" for obvious reasons. The phosphorus and sulfur compounds in that "zinc" are attracted to metal surfaces. This is particularly critical where you have a camshaft and follower interacting. The phosphorus and sulfur react with the iron that's on the camshaft and on the follower to form a sacrificial chemical coating of iron sulfide or iron phosphide. This sacrificial chemical coating is capable of withstanding the 200,000 PSI. It can handle the higher spring pressures. The zinc coatings are strong enough to keep the parts separated. Without it you would start to see scratching, scoring, and smearing of metal as a developing wear pattern. In a worst-case scenario, it could "round off" that high spot on the camshaft lobe.

Another key additive type found in many brands of motor oil is antifoam additive. This additive is typically a silicone-based product. It gets into the oil at extremely low concentrations and has the ability to burst air bubbles that are within the oil or are within the "head" of foam that's on top of the oil. If an oil formulator ignores the antifoam additive, you risk having a foaming problem. Adding some antifoam additive can give you good protection. Adding too much antifoam can aggravate the problem and *create* foam. The balance is delicate with both racing formulations and passenger car formulations.

There are several other additives including detergents, dispersants, and friction modifiers. Detergents, as anticipated, keep parts clean. Dispersants seek out the small particles and debris that find their way into the oil and keep this debris in suspension so that it will not collect to plug an oil passage nor settle out in the low spots of the engine. Instead, the debris will be filtered out or drained out at the next oil change. The friction modifier is used to help make the oil more slippery and to reduce or control the static or dynamic friction. Teams are aware of the fact that the right friction modifiers might be able to reduce engine friction, which allows the engine to create more power. This is a delicate science.

All of the components must complement each other without inhibiting the other additives from doing their job. For example, using the friction modifier "A" alone might show overall power gains. The same results might appear if the oil formulator replaces "A" with "B." But it is also possible that "A+B" in the same formulation might work perfectly together to yield a "1+1=3" result, OR the two additives might interfere with each other to yield a "1+1=0" result.

When an oil formulator selects base oils, one of the first considerations is the operating environment of the engine. A key area of stress in an engine is the journal bearings. You have a spinning shaft (either a crankshaft or a camshaft) inside of a hole (called a journal bearing). If you put a little bit of oil into that interface, the spinning shaft will help drag that oil in between the moving surfaces. As you keep supplying oil to the clearance between the rotating shaft and the bearing, the shaft actually rides on a wedge of oil. There should be no metal-on-metal contact.

If you were to measure the temperature down in the oil pan or "sump" where the dipstick is in a passenger car, you would see oil at ambient temperature at start-up, but it would slowly heat up as you cruise along. You want the oil to eventually heat up to at least 212°F. The normal combustion process, even with everything working perfectly, is going to generate moisture, which can make its way into the oil. It is simply part of the natural chemistry equation for burning fuel. By getting the oil to at least 212° you should be able to evaporate some of the moisture that is in the oil. So 212° is the minimum that everybody is striving for in the crankcase. Similarly, some unburned or partially burned fuel may also get into the oil due to natural circumstances and the 212° oil temperature will also help remove some of that from the oil.

If an engine is driven more aggressively, you will likely see more heat in the engine including the coolant as well as in the oil. We might see the oil temperature go up to 260 or 270°F depending on the make, model, airflow rate across the radiator, etc. With auto racing applications, especially in endurance races, you've got that engine producing maximum power. The hot spots in that engine are going to be *very* hot. For example, consider the point where the oil comes in contact with the pistons—specifically at the rings. The piston itself is very hot due to it being exposed to the combustion of the fuel. We could see localized oil temperatures up to 450°F around the hot spot of the rings. In some extreme racing applications, we could see oil gauge temperatures just over 300°F!

Some teams may experiment with a lower oil viscosity grade in their search for additional power. Instead of using their SAE 20W-50, for example, perhaps they go to an SAE 10W-40. By going down a viscosity grade at temperatures near operating temperature (like from xxW-50 to xxW-40), they may be able to free up some "parasitic losses." In other words, if an oil pump doesn't have to labor so hard to move the thicker oil, it may free up additional engine power, good for racing.

Sometimes teams want to preheat the oil before a race to improve flow and to thin out the oil compared to its viscosity at cooler start-up temperatures. In drag racing, for example, and in some race qualifying, teams sometimes employ a temporary external or even internal oil *heater* to get that oil thinned out to a desired viscosity so it circulates well where it needs to go and with the least parasitic drag. If you are considering preheating your oil, you might want to speak to an oil expert first—perhaps they can recommend an oil that meets your viscosity needs at start-up and at operating temperature without the need for preheating.

While some passenger car engines may hold four or five quarts or so, racing engines typically hold much more. This volume may be closer to eight quarts or twelve quarts. Twelve quarts is three *gallons.*

When the oil collects in a sump at the bottom of the engine, typically the oil pan, this is referred to as a "wet sump" system. This is used by some race teams. However, as the car hits high g-forces on turns, the oil can potentially move to one side of the sump or the other and potentially starve the oil pump of oil. Another concern in high g-force driving is that the oil level moves to a position where the rotating parts of the engine churn it up (like a horizontal blender), which can put air bubbles into the oil. Air is not a good lubricant. Air in the oil can lead to foaming, cavitation problems at the bearings, and finally, as air bubbles accumulate in the oil, the oil level rises, which puts more oil in contact with the rotating parts; this becomes a vicious cycle, putting even more air into the oil.

Other race teams will move to a dry sump system. In a dry sump system, the oil pump not only pumps the oil into the engine but it is also designed to "slurp" up oil from different strategic spots in the engine instead of waiting for the oil to drain back to the bottom of the engine. This provides several benefits. A deep oil pan no longer has to be located under the engine. Instead, a remote oil sump/reservoir can be placed somewhere else on the vehicle to collect the oil "slurped" up by the oil pump. By relocating the oil sump, the engine can sit lower in the vehicle, which can help improve overall aerodynamics and lower the car's center of gravity.

These trucks parked at the race show the two engine suppliers for Indy racing.

This is a fresh 2015 Honda engine directly from Honda Performance Development. Part of the "lease arrangement" that teams have with Honda is that one of their engineers will be present each and every time the engine is fired up. Teams are forbidden to do anything to the engine or with the engine. That said, the Honda engineer is "like part of the family" around the garage. He and the team share the same desire to win. Honda enjoys a special niche in Indy's recent history. They "stayed the course" after GM and Toyota pulled up stakes. In the years that they were the single supplier of Indy engines between 2005 and 2011, there was not a single engine-related "retirement" or blown engine for any team racing in the 500. That feat is measured against the fact that at least one such engine failure occurred in every race from the inception of the 500 in 1911 up until 2005.

# Q&A WITH BRIAN FIGG
# ENGINEER, SPEEDWAY ENGINES

*In December 2003 I interviewed Figg, and then in October 2015 I received an update (see page 27). What was originally conceived in 2003 as part of a chapter on fuel is now central to the broader topic, engines. At the time, Brian Figg, BSME, was part of Speedway Engines of Indianapolis—working hand-in-glove with General Motors Chevy Racing Division. Then, Indy race teams had three authorized racing engines to choose from—Chevy, Honda, and Toyota. Methanol had long since replaced gasoline at Indy, prompted by the catastrophic track fire at the 1964 race. Now in 2015, cars run on Ethanol E85. Figg now operates Innovative Performance Technologies (www.BuyStifflers.com). Figg jumped right into the topic of methanol vs. gasoline.*

Let's start the discussion with methanol. I have talked to people about the big crash at Indy back in '64. They appreciate the fact that a methanol fire can be put out by water. Not so when the fuel is gasoline. Gasoline is quite heavy; the vapors have a tendency to stay just above ground level. Also, the "ignition point" of gasoline is lower than that of methanol. There were a lot of devastating fires when it came to racing accidents. Gasoline vapors tend to be heavier than air. Methanol is lighter, and it evaporates a heck of a lot quicker than gas. Gas will stay "ground laden" if you will, and the "flash point" of it is lower, making it a much greater fire hazard. Flash point is the temperature at which fuels will combust.

That's what compression ratio and spark timing is all about. Compare a gasoline engine and a methanol engine. A gasoline engine, because the combustion point is so much lower, you can use—or rather, you have to use—a much lower compression ratio than you do with methanol. [For] the gasoline engine (we are talking about one that is non-turbocharged), if you are running in the neighborhood of 10:1 compression ratio, that is fairly high. That could be a normal engine, or even hot rod engine out there on the street. However a 10:1 in a methanol motor, you are really on the low side of the compression. Typically you would want at least a 13:1 for some pretty good combustion, and we have run up as high as 17 or 18:1.

## Q: WHEN YOU COMPRESS IT THAT HIGH—15:1 OR HIGHER—DO YOU WORRY THAT IT WILL COMBUST PREMATURELY?

With methanol? No—methanol is very, very stable. It is considered to have no pre-ignition qualities. With gas, because of the fact that it does go off at a lower temperature, you can get into a lot of pre-ignition. Basically, the fuel-air mixture is combusting before the spark-plug fires. This can be caused by "hot spots." This can happen because the spark plug continues to glow or stay hot after the combustion process. It might even be caused by something like a sharp edge on a piston. Pistons, if you look at them, they're always kind of "manicured"—nice and pretty. You don't see a lot of sharp edges. Sharp edges can glow as a hot spot and they'll pre-ignite the mixture before the spark tells it to. Once you do *that*, you are not having your peak pressure—where you want it in relation to the crank angle. Typically, your peak pressure (just for conversation's sake) would occur ten or twelve degrees *after top dead center.* If it is going off *before* that—before the spark plug tells it to—you may have your peak pressure *before top dead center.* Then the engine is basically fighting itself. When you do not have your "pressure peak" where you want it you're missing "optimum torque."

Pre-ignition and detonation—they're really two separate events. They can cause you to have cylinder pressure higher than that which the piston and head are designed for. This localized really high pressure can begin eroding away the pistons. I am sure you've heard people say, "We detonated a hole right through the piston!" That is exactly what can happen in a gasoline engine. Be it lower octane rating than it needed to be or higher compression than it needed to be—or *both*—it will start to have these localized high pressure areas, or "pressure waves" if you will. It will start to pick away the top of the piston and eventually just bore a hole right through it.

You have to ask what the engine is designed to do. Racing? What kind of racing? What fuel? Typically, the "lid"—the thinnest part of the piston—may only be 200/1000ths of an inch in thickness. That's a little bit less than a quarter of an inch—which sounds all well and good—but when you start looking at combustion pressures and the fact that it's running under extreme temperatures, any metal, especially aluminum, will soften up as temperature increases. Some of the stiffness and rigidity starts to go away and gradually it becomes more prone to "failure mode."

When you get pre-ignition, the "charge" is lighting sooner than it should. When that happens, you start to build up pressure. As the piston is coming up, your volume is getting smaller. Then, once the piston gets to— let's say our timing is set to 30 degrees before *top-dead center*—where the spark plug should fire but it has already *started* to burn (from whatever source), then the spark plug tells it: "Okay fire." Well, it fires and it starts another "flame front"—the "proper" flame front—propagating out from the spark plug. And at the same time, somewhere over here, in another part of the chamber, the mixture is already lit. You now have two flame fronts rushing toward each other, and as you well know, we're talking micro-seconds for all this to happen. Micro to milli—we are not talking a whole lotta time. So you have these two "pressure waves"—"flame fronts" which are in fact "pressure waves"—coming toward each other, and then they know it'll try to fatigue the piston quicker than it would any other place on the top of the piston. And if this is a repetitive process, as far as it continues to go off . . . prematurely, and . . . the spark plug fires at 30 degrees before top dead center and it continues to . . . repeat this cycle, it will definitely start to fatigue . . . a given area there on that piston; that, as well as the fact that overall cylinder pressure is higher than what the piston is designed for, plus it's starting its combustion process early. As it's trying to build pressure—typically—the air/fuel is exploding or trying to expand; at the same time the piston is coming up and it's keeping the volume. So, let's say you have a piston designed to maintain x-amount of pressure over its "fatigue life" of say six hundred miles, you greatly reduce the life of the part, which leads to premature failures.

## Q: WHEN THE PISTON IS TRAVELING UP TO TOP DEAD CENTER—WHEN IT GOES THERE—IS THAT GOING TO BE THE MAXIMUM PRESSURE?

Maximum [explosive] pressure will occur ten to twelve degrees *after* (that's where you want to have it—ten to twelve degrees after top dead center); yes, the volume is *increasing* at that point, but if everything is timed properly, you are still working on the "trailing edge" of your combustion. So you're still burning up those residual gases that are toward the perimeter of the piston area, and so that is where that *maximum* would occur. That's what you want to have timed—and all this interrelates with valve timing.

You are going to get the most *force* on your *crank* at that point— you know, ten to twelve degrees *after*. The more force on your crank, the more torque; horsepower is derived from torque. It's all interrelated. You can't have "peak pressure" at top dead center, it is kind of locked—you have got kind of a locking mechanism there. And then, it does not want to turn until you get a few degrees *past* that point—either before or past that, straight vertical—the maximum pressure that you are going to get, and the maximum work that you are gonna get out of the engine is going to be, let's say, ten degrees after your TDC.

## Q: THEY FULLY SHARE THEIR SECRETS WITH YOU?

We are contractually bound with Chevy so we can't work on any engine other than a Chevy, be it the Indy Racing League or even NASCAR. We currently work exclusively on the Gen IV IRL engine. A bit of background here. GM does not have its own R&D and Build facility at the present time. What they have done historically in all their racing venues is that they align themselves with a company that is in business already, which is considered " top of the class," if you will, and they sign up with that company to perform "development" and to do all the rebuild work and also team support for their product. That's where Speedway Engine came in. Back [in] '97 when Speedway Engines was first incorporated, we were doing Oldsmobile engines for the IRL. There were probably six to seven different engine builders doing that same engine. Basically GM would supply you with all the parts. You had the liberty to change anything, with the exception of that cylinder head and block (by rules) and then it was kind of a race between the different engine builders to see who could take this basic block and head configuration and make the most power—power and reliability. Looking at all that, the teams decided on who they wanted to do their engines. Well, we were lucky enough or good enough (depends on which way you want to look at it) lucky enough to be a leader (well, our engines won three championships—two with Panther Racing and one with Buddy Lazier at Hemelgarn [Racing]). When Toyota and Honda jumped into the program, GM needed to kind of tighten the reins a little bit and not have three or four or five different vendors out there doing their engines. [They needed to] make it more (for lack of a better term) . . . more confidential, more secretive. So they asked, "Okay, who's our best player?" So, we were fortunate to be [at the] "top of the totem pole." They aligned themselves with us. The IRL rules mandate that the engine manufacturer has to have at least one "rebuild" facility. Manufacturers obviously do a great deal of their work "in-house."

Toyota has TRD [Toyota Racing Development] out in California—that's their "in-house" rebuild facility—and they also align themselves with Penske's engine shop up in Reading, Pennsylvania. So that's their "off-site" rebuilt facility. Honda has HPD—you know, Honda Performance Development out in California, that's their "in-house." They also align themselves with Ilmor Engineering in Michigan—so that's their "off-site" as well. GM didn't have "in-house" rebuilt facility, but they (this is at the beginning of the season) aligned themselves with us.

We do anything "engine related." For example, just last week, out west we were testing different "launch controls" for leaving the pit box quickly and efficiently as well as other "yellow flag" situations; that's all stuff that was instigated and tried on behalf of GM.

Therefore, whatever we learn on an individual team's car, if it's "engine related" goes to everybody. Now, car stuff, anything that they may be doing on their car—setup-wise or aero-wise, or whatever—that is obviously their proprietary stuff. A guy could lose his job for throwing out any "team information" to another team. We're all very much aware of what's an "open book" and what isn't.

At Speedway Engines, all we do are the engines, so there are no cars here. As far as the engines go, when they leave the shop here, every engine is identical. You could take the engine out of the number four car, throw it in Cheever's car, and there's nothing different between the two.

There is basically an "engine pool" and from that pool there is a constant "back and forth" between the shop and the teams. So there isn't a group of engines—say, engines serial numbered "one through five," for example, that are exclusively "Panther team engines"—because the installation of the engine in the car, the way it's all "plumbed" hydraulically, electrically, and mechanically—like throttle cables and so on—is all designated by us. So we need to be sure that everything is 100 percent interchangeable among the engines, from team to team, and once it's installed in the car it is going to run the same.

I mean we have a package—we know what will work. Here's the engine, to its given spec (you're going to install it this way), and that's going to be, as we see it, the best—the optimum package for teams to run.

## Q: WHAT'S GOING ON IN AN ENGINE DURING THE TRANSITION FROM RACING AROUND—AT SPEED—TO PIT STOP, TO ACCELERATING BACK UP TO SPEED?

If it was an operating engine coming in off the track? You know, from the most loaded "duty cycle," which is going to be wide-open throttle, top gear, out running on the track, as it starts to come into the pits obviously, you're getting off the accelerator. So the torque will greatly diminish, and then the driver will invoke what's called the Pit Lane Cruise Button, which will take the engine from running on all eight cylinders down to, well, as many as we want it to run on (usually it's about four cylinders). So that'll keep the car and driver at the mandated sixty miles per hour going down the pit lane. At that point, no matter how hard he presses on the gas, the car won't go above sixty miles per hour. Then he comes into the pit box, and as he's coming in, he's throwing it into neutral and the engine is (well, unless the driver has happy foot) on the "idle stop." At that point it's kind of "taking a breath," saying, "Wow, that was fun, now let's go out and do that again." The oil's starting to cool down ever so slightly—the water's inclined to go up in temperature a little bit. And we have the things running really "lean"—in the pits—for fuel conservation; so it's running with just enough fuel through it so it's continuing to hit on all eight cylinders. Because once the car has come to a stop, it goes back to running on all eight cylinders because you don't want to take a chance on the thing dying.

The car comes back down off the jack. The driver drops it in gear (we have a "launch control" system in the car that's all "managed" by the Engine Management System) where he just puts his foot to the floor, holds it wide open. We limit the rpm of the engine to a given rpm number—much lower than Ten-Three [10,300 rpm]. He puts it in first gear, dumps the clutch, as the tires spin up (we allow the tires to spin up to fifty or sixty miles per hour), but the car is basically not moving at that point, since it's trying to get traction. He then pulls out onto the pit lane. By this time, the traction control has taken effect, and the traction control is calibrated such that as the car reaches, say, five, six, seven, eight miles per hour (and it knows that it's doing that by virtue of wheel speed sensors on the front tires). So once the front wheel sensors say, "Okay, we're at six, seven, eight miles per hour " ([or] whatever we decide to put in there), then it starts to cut cylinders back out the engine. It goes from eight and it starts to cut 'em down, in order to reduce torque. This allows the car to "hook up"—put spin on the tires and "launch" down the pit lane quicker! You know, the whole objective here is to get down off the jacks—launch the car out of its pit box and get back to the mandated sixty miles per hour as quickly as you possibly can. You can't go over sixty miles per hour until you reach the pit exit. Even if you're the very last pit in the train there, you still have a little way to go down the pit lane before you're at the pit exit.

Then, of course, there are other things that have a big influence—things like "How sticky is your pit box?"—you know, has it had a whole lot of rubber laid down in it? Are you getting good "grip" coming out? That's why drag racers do a "burn out" before they take off, right? They spin their tires to get 'em heated up, but also to lay down some rubber that they can use to get a good "launch."

You can hear it. If you have a "tuned ear" you hear a big difference between an engine just sitting there and one just "burnin'" out of the pit on high rpm—on the 10,300 rev limiter—versus one that . . . I mean you could hear the engine come down, it will actually kind of "spin up." [*imitates sound*]

In our testing in the southwest we worked a little bit on soft rev limiter control. Of course you well know that we're mandated to 10.300 rpm maximum rev limit—and that is controlled (primarily) by our engine control system; but the IRL also has a sensor that they put on the engine that not only *monitors* our rpm but also will *cut* it if it gets above 10,300 rpm. So the best thing for us is to take our system and tune it electronically to let the engine run as *close* to 10,300 as it possibly can without going over. Once it goes over, the IRL rev limiter cuts the ignition, and it's basically like driving around and cutting the ignition switch off. It's not good, and the engine doesn't like it. It's quite a penalty "time wise." It also upsets the chassis; if a guy's going into turn one, and he's on the IRL rev limiter, it's really harsh. It cuts the engine *low*; and when you cut the power in the *corner*, you're basically *unloading* the chassis. So then the thing will feel really "twitchy," like it's trying to get away from him. So you have to be very careful. That's a subtlety that we work a *lot* on. That can make a *huge* difference in performance.

Some of the drivers are really good—if they have the feel. Say they're in sixth gear and they're pulling up behind somebody—getting ready to pass 'em, going into a corner—and they feel that the car's going to accelerate enough to activate the IRL rev limiter. They'll actually back off the throttle a little bit (Sam Hornish used to be *great* at that; he'd back off the throttle just a little bit) to keep it out of the IRL rev limiter, but still have enough momentum to slingshot past the guy in front of him. As soon as he gets *out* of that guy's draft and gets up alongside of him he can go back to "full throttle" and you haven't been hit with any "rev limitations."

*Brian Figg responded to some 2015 fact checking:*

### Q: WHEN WE WERE SPEAKING BACK IN 2003, INDYCARS WERE RUNNING METHANOL.

Yes, up until 2006 when the fuel was changed to contain a 10 percent blend of ethanol. The following year, 2007, the series switched to running 100 percent ethanol, and they still do today.

### Q: I'VE BEEN TOLD THAT THE SWITCH TO ETHANOL WAS A MARKETING DECISION. THE ETHANOL PEOPLE ARRANGED A LICENSING AGREEMENT IN ORDER TO BECOME THE OFFICIAL FUEL.

Correct.

### Q: I'VE HEARD IT SAID THAT THE CHEMISTRY WAS SO SIMILAR BETWEEN METHANOL AND ETHANOL THAT IT WAS NO BIG DEAL FOR THE ENGINE MANUFACTURES TO SWITCH OVER.

This was only true for the 2006 season, when a 10 percent blend was used, as stated above. Subtle changes were made in the engine control software to adjust for the small amount of ethanol being used with the methanol. However, there was a fair amount of development that was done to the engines to prepare them for the switch to 100 percent ethanol for the 2007 season, mainly in fuel delivery and control software areas.

### Q: METHANOL RACING FUEL HAS AN OCTANE RATING CLOSE TO 110?

Yes, the RON [Research Octane Nating] is 110 for methanol and 116 for ethanol.

### Q: THEY CURRENTLY POINT OUT THAT IT'S E85. WHAT'S THAT?

I believe the IRL continues to market the fact they are running "100 percent fuel grade ethanol." The "E" in E85 denotes "ethanol," while the "85" refers to the fact it is 85 percent pure ethanol, while the remaining 15 percent comprises gasoline and other additives.

As a side note, "100 percent fuel grade ethanol" is not 100 percent ethanol. The government restricts the sale of 100 percent pure ethanol because it can be consumed by humans; therefore two percent gasoline is added to denature it.

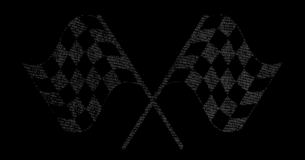

# 3
# TRANSMISSION

**THE FRENCH** term for transmission is *boite de vitesse*—literally "box of speed." It is a deceptively simple mechanical device. The XTRAC gearbox (Indy's exclusive supplier) has the technology to allow teams to "mix and match" actual gears to arrive at specific "gear ratios" in the frantic hours before a race. Such decisions are made based on weather, temperature, track conditions, and the ethereal driver and team "instinct." If they are prescient in their "setup" they may achieve faster acceleration for passing or "restarts" or allow their driver to get out front, stay there, and run cool all day. Individual gears are about the size of a hockey puck but represent the ultimate in metallurgy and mechanical precision. The engine is capable of making 15,000 revolutions per minute. The "tranny" is the crucial device that determines how that potential is delivered to the rear wheels.

A gear cluster they refer to as the "Christmas tree" sits ready to be "swapped out" at any moment if different gearing is called for. Beside it is a pre-measured jug of its lubricant.

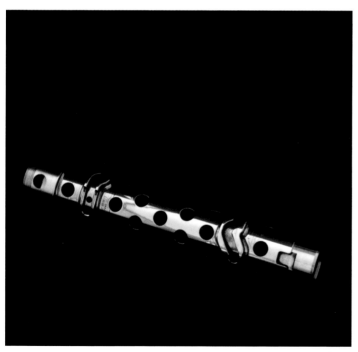

# Q&A WITH ANDREW HEARD
# ENGINEER, XTRAC

*Xtrac engineers have been very supportive and patient over the years. Here, during Race Week in 2015, Andrew Heard proudly describes what his gearbox does to keep a race car moving.*

## Q: HOW QUICKLY CAN A RACE TEAM SWAP GEARS IN AND OUT OF YOUR TRANSMISSION?

The way it is now they take the rear attenuator off the car, which includes the whole rear bumper assembly; then there are about 10 nuts to "induce." Then you pull the whole cluster out and put another one in. There was a team in about 2004. They had the gearing all wrong. The race had begun and they saw that they were set up with the wrong gearing. So during a "yellow flag" period they actually elected to change the gears. Still, they finished the race on the lead lap. They got it done very quickly. We call it a "cassette cluster," so the whole lot is ready to go—the shafts, the gear change barrel, the selector mechanism—everything like that. It's an assembly; you're basically taking that whole lot out of the casings and pulling out a lot of the guts of the gearbox, then putting in a different cassette, which is set up with different gearing.

It wasn't that it was a fault on the mechanics' side; it's just that the engineers had predicted things slightly off. The weather was different from what they expected. They could have run around uncompetitively for three hours, so they decided to do something about it.

## Q: WHAT DO YOU THINK OF THE FACT THAT SOME TEAMS USE THINNER OILS, AND LESS OF IT, FOR QUALIFYING?

They'll do anything that they can, balancing costs, of course, for potential failure. It's reduced drag, basically. It could also mean reduced life of the parts. For example, if you took it to the extreme of "not running oil" after a qualifying run, it wouldn't break, but there would be parts in there that . . . would have to be changed out. Incidentally, I'm not sure that there'd be an efficiency gained by

running "dry." When you do a tuning change to the transmission, most of the lubricant comes out at that time. Depending on the team's schedule and when they last changed it, they may replace that oil with fresh oil, or just put the same oil back in. It's really down to the team's "procedure." The oil usually comes out very clean.

## ON PARTS LIFE, BAR CODES, ETC.:

We laser etch part numbers on all the parts, as well as their "ratios." The other key parts are "serialized" for quick identification. The teams may elect to vibro etch their own numbers on there as well.

## Q: CAN TEAMS ELECT TO NOT INSTALL THE "REVERSE GEAR" DURING QUALIFYING OR FOR THE RACE AT INDY?

No you cannot. You're not allowed to do that. That used to be an option on the older model gear box. They're not allowing that option anymore. You have to have a reverse in at all times.

Historically, an "ovals only" transmission did not have reverse. So there's an argument: "You don't need a reverse" or "we don't have to have it in there". . . but now these cars switch between ovals, road courses, and street courses, and the gearbox has reverse. We're trying to prevent the teams from . . . applying too much "specialization" of the gearbox for one particular race "type." So there's no reason that you couldn't take the transmission from here at Indy, go straight to the street course in Detroit, and run it. They'd just change the spools, the differential, and the ratios.

## ON THE NICE, CONICAL STACK OF GEARS:

The "Christmas tree" is what we call it. It was the same transmission between 2003 and 2011; now it's the same from 2012 to the present.

## ON PADDLE SHIFTING:

On the new car, in addition to paddle shifting they also have a hand-operated clutch. There's no clutch pedal and no foot pedal. The clutch is the lower paddle on the backside of the steering wheel. I think they have one on each side, but it's the lower-down position. We don't do the clutch actuation parts. That's an AC Racing assembly. It's a pneumatic system. We do supply the "air" to the clutch; the same with the gearshift—that's pneumatic as well. Because we already have the pneumatic gearshift system on the car, rather than putting another compressor on the car, they "T" in and we supply the air.

The driver actuates the clutch on a handle, very similar to what you have on a motorcycle. He pulls the other paddle for an "up shift." It goes into gear and he just lets the clutch go; from then until he needs to stop again, he won't use the clutch. He'll just shift up and down using the paddles. Yes, it's still possible for him to "stall it," but he won't "grind" the gears. There's a very small chance it won't "go in" (engage); then, all he has to do is "try again," but it won't grind the gears. His need for the clutch is to engage the "drive." It's more like a safety feature. It won't allow him to the engage gear unless he pulls the clutch in.

The driver can then freely go up and down through the gears without using the clutch. He can even go into the lowest gear as they might do on a street course—such as the final turn at Long Beach, for example. On some tracks they may or may not use first gear, except on the pit lane.

An interesting thing running here at Indy, it is a six-speed gearbox. You'll find that first, second, and third have a very wide "spread" on them, and they'll run here with fourth, fifth, and sixth very "close"; that is, with very little difference between them. Depending on the speed and depending on the condition of the track, they could be running in any one of those "top" gears. If they're going to do a "229" lap they're probably going to be in sixth gear. They're going to do a "226," say? They might be in fourth gear—it's a "sequential shift transmission," so he'd have to go through fifth gear on his way down to the fourth. If there's a lot of wind, they may be in sixth gear with a tailwind and switch to fifth gear as they come around into the headwind. With most of the "tuning" here . . . they pretty much know what they're going to need to start on pit lane with and they all know pretty

much what speed they've got to get up to, so they do those "splits" to get the best acceleration to that speed. Then they'll be playing with fourth, fifth, and sixth gears, tuning them . . . for qualifying they can be looking for a change as small as sixty rpm. They're all just trying to make sure that they're running at the peak power of the engine for the speed they believe the car is going to run. There's a "rev limiter" at 12,000 rpm. I can't tell you exact torque, but I reckon that these cars are producing power right up to that limit.

I haven't heard the cars running up into that limiter lately. Actually, they don't want to get up "into" the limiter. Back a few years ago, you'd hear them all the way from the start/finish line into the first corner "on the limiter," because that's where the peak power was— right on the limiter. You'd have to get your gearing "just right." What the manufacturers did (this is way back) . . . there had been a "series mandated hard limiter," which really takes away power. So the manufacturers worked on making a "soft limiter" just below the hard limiter so that you couldn't even get "into" the hard limiter. If you get into the hard limiter you're really losing a lot of forward motion.

## BACK TO SIXTY RPM:

Yes, we're talking engine rpm. So, for example, if their peak power is at (figuring the limiter at 12,000) 11,800 rpm, and the speed of the track is only allowing them to 11,700, they'll change the gearing so that they're running in the peak power of the engine.

Wind is the biggest factor in gearbox tuning. Second would be the weather. Then, track temperature. The hotter the track gets . . . it's a curve, isn't it? If the actual track temperature is "too cold," then there's not as much grip. Tires don't get as much heat, so there's not as much grip. The air temperature . . . with "colder" air temperature you get more engine power. Also if they can only "trim" the car so much you get more "aero drag" as well. These cars, they can trim them more anyway, so . . . [laughs]

There's a lot of different factors that come into it—air density, humidity, barometric pressure . . . the discussion could continue.

# 4
# TIRES AND THE TRACK

IN RACING, the spot where the rubber meets the road is called the *contact patch*. When a car is at full speed the temperature there could boil water (212 degrees Fahrenheit). Firestone is the exclusive supplier of racing tires. Their research, development, and manufacturing seek to create a tire composition that will stand up to the rigors of a particular track surface on a particular day. A tire raced at Indianapolis, Indiana, at the end of May will require a different "formula" or "composition" from one that will compete at (say) Phoenix International Raceway in August.

The tire plus the forged aluminum rim weigh, in total, less than thirty pounds. The walls and surface (Indy tires have no tread) are about the thickness of "three credit cards," according to Firestone's Dale Harrigle (see sidebar, page 46). Generally they are expected to last as long as one full tank of fuel. That's currently 18.5 gallons. Depending on racing conditions, cars can get between three and four MPG. No driver wishes to "hit the brakes" during a race except when coming in for a pit stop. Doing so—perhaps to avoid a collision—can "flat spot" a tire. A good quantity of rubber is usually "scrubbed off" during extreme braking so that performance is severely compromised and a "tire change pit stop" will be scheduled. Tweaking the tire pressure at each of the "four corners" of the race car during pit stops can dramatically affect handling and performance.

Both Buddy Rice and Dario Franchitti garnered Indy victories in races that ended in conditions such as we see here. There is no windshield in "open wheel" racing, only a helmet with a visor. Harsh conditions such as these will bring out the red flag. Every team has access to sophisticated weather maps and data to work out their strategy. Choosing to "pit"—or not—can be the deciding factor in such situations.

There is a small canister in the six o'clock position of this OZ Racing forged aluminum wheel (rim). As soon as the wheel and its mounted tire are "rolling" it transmits a steady stream of temperature and pressure data back to the race team. The wheels are owned by the teams. The tires are "leased" from Firestone.

The race car covers quite a bit of ground around Gasoline Alley and out to the track. It's generally towed or pushed without starting the engine. It visits the fuel area, the inspection plate, and is brought from garage to track for qualifying, practice, and racing. You'll frequently see that it's shod with these "tow tires" also known as "roll-arounds"; it's also said that they are "gifts" from Firestone. They raise the car an inch or so higher than the smooth racing tires, which tends to protect the precise undercarriage and "tunnels" from bumps in the pavement and little pieces of debris.

"Part life" is a major concern for a race team. Where, when, how, and how long a part has been used is carefully monitored. Tires and wheels are no exception. One of the empty wheels in the stack is marked to show that it was involved in an accident (above, left). If a tire has seen limited use for practice, testing, or qualifying, it may be dismounted and put aside for future use. The direction it "turned" is marked with crayon, as is its position on the car.

"Laying down a little rubber" in the pit box is desirable for a couple of reasons (besides beating a competitor to the "pit out"). It warms the new tire, which will be achieving full, desirable pressure during the first lap. The sticky black marks it leaves in the box will add grip for a quicker "launch" in subsequent pit stops.

As cars are racing "nose to tail," the following car is lashed with the sticky black threads of rubber that fly from the car ahead. The "tear off" plastic sheets a driver will have on his visor will clear his vision for a while. If the car is mounted with an on-board camera which has a tiny aperture, one glob of hot tire compound can totally block the view until its clearing mechanism is activated.

**GASOLINE ALLEY** - With our back to the actual garage area, this is the main thoroughfare for teams to shuttle race cars, equipment, and tire wagons from their assigned garages to the pit or actual racetrack. When this area is thronged with people (nearly always), a lively cadre of "yellow shirts" pipe up with their sharp whistles to create the "right of way" for the racers.

The tire allotment for each race is a set number. You're not allowed to "buy more" if your supply is low. Hence, decisions are made about which tires get used—and when. A set used for a short practice session may have a lot of "laps" left in them. If they've picked up a lot of "marbles"— the gummy debris that litters the track surface (plus cigarette butts, pebbles, and other gunk)—the tires can be warmed, scraped clean, remounted, and reinflated.

Indy has a century of experience hosting races and the efficient layout is a marvel to see. Haulers (or transporters) have an assigned parking area. Garage space is equitably meted out. Hospitality tents for teams, suppliers, and sponsors have their separate areas. Fueling has a strict protocol and takes place in an assigned area. All this happens with the understanding that the race fan wishes to experience all the "behind the scenes" activity. If your ticket purchase offers an optional "garage-area access" it's a good value.

The starter's "flag stand" looms over the track above the start/finish line. One of the traditional logos used by the Speedway is topped with the seven signal flags used at Indy to communicate with the racers. The simplified meaning of each is:

Green flag: Indicates the start of the race. It may be used again as a signal to "resume racing" after a caution period.

Yellow flag: Displayed to order the cars to reduce speed and refrain from passing due to "unsafe condition on the track." The cause could be an accident, debris on the track, disabled vehicle, light rain, or similar conditions.

Red flag: Indicates that the race has stopped. Drivers should return to the pits. Severe weather or a major accident can bring this flag out.

Blue flag: Seldom seen, but it can be displayed to a single car to indicate that a faster car is approaching and "racing courtesy" dictates that he move over and allow a pass.

Black flag: When directed at a specific car is a direct order to report to the pit for a penalty to be assessed or to attend to a repair on the vehicle.

White flag: Comes out to indicate to the lead driver(s) that one lap remains.

The checkered flag tells the lead driver that he has won the race; it's continually waved to indicate to the following drivers that they should slow down, as the race has ended.

This is among the most thrilling moments in sports. The pace car has completed his task and left matters to the driver sitting on the "pole" and the starter on the flag stand. If the green flag drops, there are just 200 laps remaining.

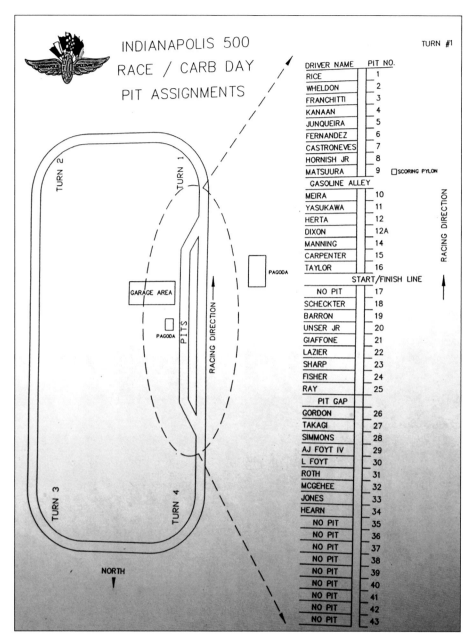

INDIANAPOLIS 500
RACE / CARB DAY
PIT ASSIGNMENTS

TURN #1

| DRIVER NAME | PIT NO. |
|---|---|
| RICE | 1 |
| WHELDON | 2 |
| FRANCHITTI | 3 |
| KANAAN | 4 |
| JUNQUEIRA | 5 |
| FERNANDEZ | 6 |
| CASTRONEVES | 7 |
| HORNISH JR | 8 |
| MATSUURA | 9 |
| GASOLINE ALLEY | |
| MEIRA | 10 |
| YASUKAWA | 11 |
| HERTA | 12 |
| DIXON | 12A |
| MANNING | 14 |
| CARPENTER | 15 |
| TAYLOR | 16 |
| START/FINISH LINE | |
| NO PIT | 17 |
| SCHECKTER | 18 |
| BARRON | 19 |
| UNSER JR | 20 |
| GIAFFONE | 21 |
| LAZIER | 22 |
| SHARP | 23 |
| FISHER | 24 |
| RAY | 25 |
| PIT GAP | |
| GORDON | 26 |
| TAKAGI | 27 |
| SIMMONS | 28 |
| AJ FOYT IV | 29 |
| L FOYT | 30 |
| ROTH | 31 |
| MCGEHEE | 32 |
| JONES | 33 |
| HEARN | 34 |
| NO PIT | 35 |
| NO PIT | 36 |
| NO PIT | 37 |
| NO PIT | 38 |
| NO PIT | 39 |
| NO PIT | 40 |
| NO PIT | 41 |
| NO PIT | 42 |
| NO PIT | 43 |

□ SCORING PYLON

RACING DIRECTION

TURN 2   TURN 1

GARAGE AREA

PAGODA

PITS

PAGODA

RACING DIRECTION

TURN 3   TURN 4

NORTH

Indianapolis Motor Speedway's press handout showing pit assignments for both Carb Day and Race Day. Qualification is obviously finished with Buddy Rice on the pole and Dan Wheldon and Dario Franchitti sharing the front row. The schematic depiction of the track is useful "fiction" to help visualize the path cars would take from turn four into the pits and then out again. In a true-to-scale map of the 2½-mile course the racing surface would be rendered as a spaghetti-thin line.

Consider the "honor" of occupying pit #1 (etc.) according to the fastest qualifying times. Mathematically all racers will cover the exact same distance in both racing and pitting. It's felt that the main advantage to a forward pit assignment occurs when a cluster of cars pit at the same time. The low pit numbers encounter less chaos from cars exiting their pit.

The red flag with the yellow "X" is a "local" flag not displayed from the starter's stand. It indicates that the pits are closed.

This is one team's signage resting against the wall in the "trough" between pit lane and the front straightaway. Messages are short, sweet, and sometimes cryptic. Radio communication can sometimes be faulty, and getting the message across as the driver whizzes by can be vital.

Drivers and teams alike are well schooled in both "general rules" and "rules of the day" as might be outlined in the private "drivers meeting." This gentleman is one of many who see that things run smoothly and fairly.

When most of the field elects to pit on the same lap it sets off "orderly chaos" along pit lane (also known as pit road). The cars will fastidiously observe the 60 mph limit up to their assigned pit. Leaving the pit after fueling and tires are changed is a different, highly charged situation (the speed limit is still in effect). A speedy stop and a sharp exit from the box can mean advancing in position. It's just as satisfying as passing those cars on the racing surface.

Race Week is peppered with events that enhance the "racing experience." Here, vintage vehicles (sometimes piloted by vintage drivers) course the Speedway. The Speedway museum proudly keeps most of their vast collection in "racing trim." The sounds alone are stirring. Although frequently billed as a "competition," all know and respect these valuable artifacts of racing's past.

# Q&A WITH DALE HARRIGLE
# TIRE ENGINEER, FIRESTONE

*Dale Harrigle took time out of his busy Race Week (specifically Carb Day 2015) to share some specifics on racing with Firestone tires.*

The teams lease the tires from Firestone. There are probes inside the rim [to calculate] tire pressure, temperature, and infrared temperature. It's in one unit, but it measures three separate things. Some teams balance the rim to compensate for that so the rim is perfect. Other teams allow the rim to be out of balance and we mate it with an out of balance tire and we balance the assembly. It depends on the team.

There are teams that make sure that the wheel itself is perfectly balanced, so they put on the sensor and then they put on additional weight until the wheel is perfectly balanced. There're other teams that allow the wheel to be out of balance. Then the tire is out of balance, [and] when you mate the two together you just balance the result.

Between those two systems I would actually leave the rim out of balance. It sounds strange, but it can lead to putting less weight on the assembly. The less weight you put on there—[in] my mind—the better.

The tire's about 3/32nds of an inch thick. We'd say "three credit cards thick."

A stack of coins "three quarters high" would be a little bit too thick.

The tire is smooth, but as we watch qualifying laps, the drivers leave pit lane and their first lap can be in excess of 200 miles an hour. So the tire has pretty good grip, even when cold. Tires grip the surface in two ways. There is an adhesive component to the grip, so they do actually get sticky and in a sense it would grip a sheet of glass. It adheres to the surface. There is also what we call a "mechanical keying" where the rubber is soft and actually blends into [the pavement, in a way]; you know pavement is not perfectly smooth. Pavement has little areas of void and whatnot. The tire actually gets down in there and keys into the void. That's the second component of grip for a tire. The tires are

anywhere from eight to ten psi—"low" [pressure] when the car leaves the pits—and they gain about that much pressure running. A day like today we would inflate to the mid-forties on the right-hand side and around thirty psi on the left. For road courses we would inflate with considerably less pressure. It would be about twenty-five psi in the front and twenty in the rear. At an oval like Indy, the outside tire is about fifteen psi higher than the left (inside) tire.

## Q: HOW MANY INGREDIENTS IN A TIRE COMPOSITION?

There are certainly hundreds of ingredients. I can't give you an exact number, but it's certainly in the hundreds. Every tire on the car is a unique size. The right front is a little larger [in outside diameter] than the left front.

Let's look at the number 14–5/27–0/R15.

The 14.5 is the width in inches.

The 27–0 is the outside diameter.

R15? R is a radial tire with a 15-inch bead.

When the tires are built they are built for a specific position on the car. The stagger here at Indy is about .32 of an inch. So the right rear is about three-tenths of an inch larger [in] diameter than the left rear. [The] stagger here in Indianapolis is actually low when compared to that of other tracks. The teams don't get to decide. We provide each team with a specific "stagger number." There is no option. They get what we supply.

There is absolutely a different tire formula for each specific track. Texas and Chicagoland would run about a ".45 stagger," so the right rear is about four-tenths of an inch larger. Phoenix specs are even larger. It's more like half an inch (right rear) larger than the left rear. Stagger is based on the geometry of the track—the bank angle and the radius of the corners.

We also adjust the compounds of the tire based on the expected ambient weather and temperature . . . in fact, a number of years ago when the series finale was at Chicagoland we actually switched specs because we had expected ninety-degree days. Then the weather was forecast for the sixties. We actually brought different tires to that race based on the weather forecast.

All these tires are manufactured in Akron, Ohio, in the plant that Harvey Firestone founded in 1911. There are belts underneath the tread. We don't reveal specific composition or what kinds of material we use . . . but you can actually wear the tread to the point that the belt is exposed. Obviously, at that point the tire has very little grip.

## Q: YOU MOUNT FRESH TIRES ON THE TEAMS' METAL WHEELS (RIMS), BALANCE THEM, AND FILL THEM WITH AIR. MANY TEAMS THEN DEFLATE THE AIR AND REFILL WITH NITROGEN. WHY IS THAT?

The nitrogen is dry. It has no humidity in it. The humidity in the air causes a larger pressure change with temperature change than "dry air" does. So they use nitrogen to add "very little to no moisture" in the tire. If you threw a little bit of water in the tire and had a lot of moisture the pressure will change drastically as the temperature changed.

The tires operate somewhere around 200°; 180° to 220°—somewhere in that range.

Fronts weigh about eighteen pounds.

Rears weigh about twenty pounds.

The wheel? Fronts I think are about eleven pounds. The rear wheels are probably a pound or two more. I believe they are an aluminum/magnesium alloy. Here at Indy, for qualifying, practice, and the Indy 500 race the team gets thirty-six sets of tires. That's thirty-six *sets*—not thirty-six tires.

## Q: HOW MUCH PRACTICE TIME IS ALLOTTED TO THE TEAMS?

For Indy it's something like six hours a day for five days: or thirty hours.

## Q: I'VE HEARD COST FIGURES OF $700 PER TIRE. IS THAT ACCURATE?

Sorry, we don't comment on price.

# 5
# TECH/ELECTRONICS/TELEMETRY

THE DRIVER will flick his eyes away from the racecourse for but a fraction of a second. The center of his small steering wheel is clustered with lights, indicators, buttons, and switches. He can see his speed, engine rpm, key temperatures, fuel level, and a battery of other information. His gloved hands (in concert with foot controls) allow him to maintain his grip on the wheel as he uses "paddle shifting" to move up or down through the sequence of gears.

A simple push of a button sets in motion a program that efficiently decelerates the car (for exiting to pit road) so that the speed limit there (sixty mph) will be maintained. In fact, *hundreds* of bits of vital information are transmitted wirelessly—in real time—from the car back to the engineers and team leaders sitting high atop the "war wagon" in his pit area. Though the technology exists for it, rules prohibit "tuning" the car as it flies around the track. The information gleaned from this telemetry can be used to manually make certain adjustments only during a pit stop. The teams will already have a "heads up" if something in the car needs attention. The crew can know the internal pressure and internal temperature of each tire. They will know if a driver is "lifting" (easing off the gas pedal), even if the driver insists on the radio "I'm running flat-out through all four corners." They can know temperatures and pressures at multiple points

in the engine and exhaust. They can measure combustion efficiency and the general "balance" of the race car. The telemetry plays a huge part in "pit strategy," where a team may risk "staying out one more lap" if worn tires and low fuel permit. Pitting "out of sequence" from the bulk of the pack can make for an easier time getting onto—and off of—pit road.

In manufacturing and marketing, the term "feature creep" is a familiar one, especially when applied to modern technology. The beta version or prototype often lacks innovations and improvements that will come with the 1.0, 1.1, 2.0 (and so on) versions. The concept of feature creep is quite germane to the development of Indy race cars, the racetrack, and the sport in general. It is not safety alone that prompts change and innovation. Competition, the "need for speed," and the "racer's edge" are powerful forces as well. In looking at a century of racing at Indianapolis there are *decades* where nothing changes very much. Then single innovations—the rear engine, the aerodynamic "wing," or innovative tire composition—can revolutionize the sport. It causes the teams to scrap the equipment they *were* using and experiment with the *new*. A team with sponsors with deep pockets can become quite innovative. Some little trick that saves a half second on each lap . . . well, there are 200 laps in the 500-mile race.

*Everybody* measures! The teams twist and tweak, looking for a little more performance. The IRL has the tech plate. They have all their little jigs and guides and gauges to assure that it's all done within the rules.

The Indy tech plate has gone high-tech. Online Resources has a presence here at the 2015 race as a sanctioned part of the IRL's technical inspection. It doesn't replace the weighing, poking, prodding, and measuring that takes place surrounding a race, but it does add an interesting capability. The curvaceous part or assembly to be tested is simply placed on the work stand, which has a small pair of rigid globes attached to the top surface. One of the "bifocal" lasers is shown on the tripod in the foreground. The laser "orients" itself by recognizing the globes and proceeds to "scan" the part in three dimensions. The measurements are compared with the "ideal" and "IRL specified" CAD image. The entire surface of a large piece such as the one shown can be completed in about fifteen minutes. They employ a decidedly low-tech method in certain circumstances. If the piece being verified is carbon fiber as it comes from the factory, the laser can "see" into the top film of resin and is uncertain where the true surface begins. A light dusting with talcum powder knocks the shine off the glossy black skin and affords the laser a proper view. The same technology has been used to harmlessly "map" valuable sculptural works of art in all dimensions, which can then be used to "print" an ultra-accurate facsimile.

"Data Geek" is an affectionate name for certain engineers on a race team. Technology permeates every corner of the sport. Pi Research was an early entrant in racing. Data could be gathered from wind tunnel testing, engine performance, tire temperature and pressure, driver performance, and endlessly tweaked and massaged in pursuit of "the racer's edge."

Timing and scoring was once the domain of a human with binoculars, clipboard, and stopwatch. Keeping track of the entire field, leads, passes, fouls, infractions, and lap counts was tedious and thankless. Still, disputes arise, but the electronic record of each vehicle's behavior is quite precise.

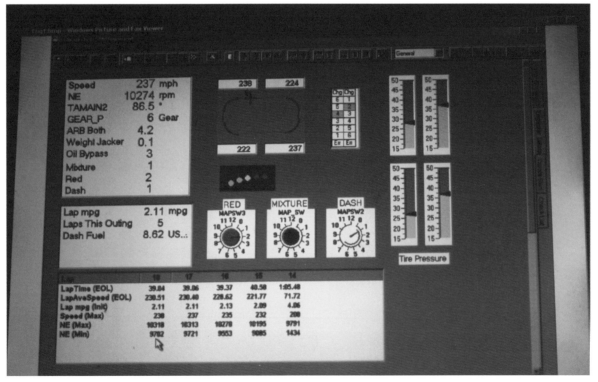

Ascertaining the ride height.

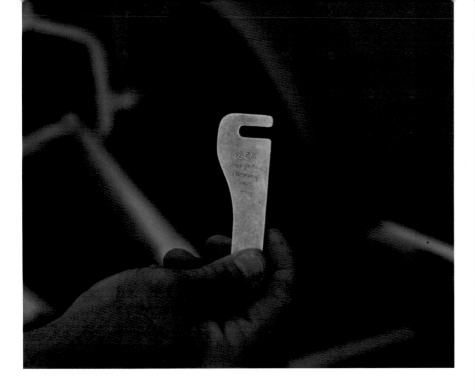

The rear wing was required to conform to this shape.

One of four weight scales—one for each wheel.

Tech plate ready for the next inspection.

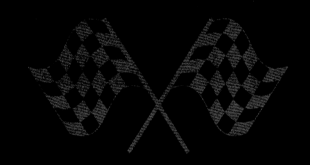

# 6
# SAFETY

**THERE WAS** a famous highway safety slogan years ago: "The speed that thrills is the speed that kills." The dangers in racing are enormous. Though *every* effort is made to keep drivers, fans, and officials safe, collision and fire are the main concerns.

The driver sits low in his "open-wheeled" cockpit. Without a windshield his helmet and visor are all that protect him in a crash. The hope is that rescue workers will be on the scene—fire extinguisher in hand—within (perhaps) fifteen seconds of a crash. A Nomex racing suit and flameproof socks, underwear, gloves, and head balaclava protect every possible square inch of skin. The first responders (also dressed protectively) attending to a wrecked driver are among the most highly trained EMTs. The track surgeon is a minute away. If the injured needs more extensive care, a medevac helicopter can be dispatched from a pad in the infield to a downtown Indianapolis hospital in minutes. The race control officials have drummed into the drivers' heads the best course of action when the yellow caution light indicates a problem somewhere on the huge track. There is a series of such lights throughout the racecourse. A light on the driver's instrument array will pop on simultaneously.

The race car itself is built to provide protection for the driver. The driver is strapped into the car with a six-point harness around the waist and chest. The rear of the seat is honeycombed material that crunches to absorb deadly g-forces that occur when the car (most frequently) slides backward into the outside wall. The weight of the rear-mounted engine tends to send an out-of-control car on that tail-first trajectory. The carbon fiber "skin" of the race car is designed to go away upon severe impact. It is one of many impact-absorbing features. The driver will still be strapped into the "tub"—a precisely welded tubular steel cocoon.

Even the walls of the racecourse have safety innovations. SAFER barriers (deployed mostly at the turns) *look* rigid but have a sandwich of Styrofoam slabs behind a broad steel strap. They are meant to absorb g-forces but not have so much springiness that they pinball the car back into the line of traffic.

Racing has become far safer in recent times, but the loss of life at Indy over the decades is chilling. The number of fatalities (which approaches sixty) includes losses at the race itself, as well as those during time trials, testing, and practice. It includes the death of crewmembers, spectators, and track workers. Each tragedy, of course, is followed by acute analysis and introspection. Yet the inherent danger associated with the sport never goes away. That said, many safety (and performance) innovations that eventually find their way into the street car were first applied to race cars.

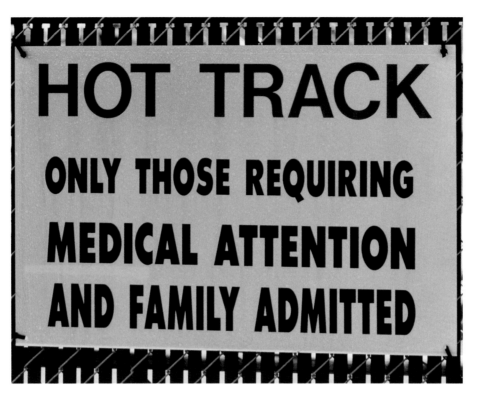

# HOT TRACK

## ONLY THOSE REQUIRING MEDICAL ATTENTION AND FAMILY ADMITTED

Safety is obviously on the mind of Ryan Briscoe as he "saddles up" for his first run in the Schmidt-Peterson #5 car prior to the 2015 race. He has been named as replacement driver for James Hinchcliffe, who had a serious accident just days before. Here we can see the fireproof "head sock" as well as the head and neck protector that all racers wear.

The Holmatro safety truck is packed with every conceivable bit of gear likely to be required if driver extraction from a mangled race car is required.

Trackside safety workers, EMTs, and other first responders can frequently be seen at the edge of the track during the warmup laps before a race. It's a very emotional "salute" that indicates the message: Race safe, but if you need our help, we're right here.

**Yellow Caution Light:**

When "race control" activates the caution button they wish "all concerned" to be alerted simultaneously. A yellow light pops on in the center of every driver's steering wheel. Lights such as shown here flash to life as well. The message is reinforced with a brisk voice in the driver's earpiece: "Caution! Accident, Turn Three!"  Their own spotter can then chime in with directions such as "stay low" or "go high" to permit the driver to "steer clear."

A high-speed wreck leaves a "debris field" of pieces both large and small. Oily residue must be cleaned as well. Even the "oil dry" (some say it's the same as kitty litter) must be completely swept clean before racing can resume.

There is protocol attached to every aspect of fuel handling. Each team has a fuel tank at their pit which is capable of holding a few hundred gallons. The fire department is very much aware of times when the hose, which connects from the tank to the race car, may contain a bit of fuel. Here they are clearing the contents of the hose into a small collection tank.

ColdFire cannisters of this fire suppressant are seen everywhere. It can be handheld and used to douse a flame and inhibit further combustion. There are other chemical suppressants with the ability to instantly stop combustion (such as Halon), but care is taken to insure that the driver is not inhaling something noxious.

Jeff Troyer of Simpson performs scheduled maintenance on Al Jr.'s racing helmet. If the visor is pitted it will be replaced. A supply of clear plastic "tear offs" can be seen. The circular tab is just to the right of the Simpson sticker. A stack of about seven can be mounted. Some drivers prefer that the tabs be alternately set with the tab first on one side and then the other, for sure grip with a heavily gloved hand. A drinking tube and the connection for the driver's earbuds and microphone can be seen.

## The Hot Parts Bucket:

Tranny mechanics know how hot the internal parts can be! In an accident, those red-hot pieces of metal can be scattered across the racetrack. Sometimes sweeping the debris into a pail causes the part to melt a hole right through the thick plastic. Therefore some handy "fabricator" welded up this square bucket from quarter-inch steel plate.

**Catch Fence:** Traditionally, the heavy steel fencing, which curves inward toward the track, was designed to keep errant wheels, tires, and other parts of the racer from flying into the grandstands. Today, a wrecked race car is designed to fracture "safely" in the act of absorbing the energy of impact. As this happens, a sturdy tether system acts to keep all the large parts like wheels, engine, and transmission from scattering.

A severe problem happens when air gets under a wrecked race car and sends it "airborne." In that case the catch fence acts to keep both car and driver on the track.

This is the "state of the art" (2014) safety gear for the fueler who goes "over the wall."

**Fuel Filling Probe:**
Here the technology has been "borrowed" from the military. The probe is alleged to be identical to the "nozzle" used in air-to-air refueling of the F-16 fighter aircraft.

Jet blower for track drying.

**Accident Data Recorder:**
The "black box" will store very precise data generated before, during, and after a car is involved in an accident.

**Smaller Jet Dryer:** Rain before or during a race may have subsided, but the track is still dangerously moist. The huge paying crowd can't be expected to wait until the sun dries the track. Actual jet aircraft engines have been mounted in such a way that the hot exhaust is directed down at the wet surface.

**Safer Barrier (Steel and Foam Energy Reduction Barrier):** Hitting a rigid concrete wall at racing speeds is abrupt and dangerous. As early as 1998 an attempt was made to lessen the force of such impact. The result is the Safer Barrier—now widely used in tracks of all kinds.

Initial tests were installed in "outside walls," usually at corners where losing control could send the car into the wall. The first point of contact was a steel band that to the naked eye resembled the original concrete wall. A thick stack of Styrofoam-type bricks just behind the steel band was designed to sacrificially crumble, absorbing a great deal of the energy. The steel band was held in place by mooring cables until a fresh sandwich of energy-absorbing foam bricks was installed. Engineers carefully avoided bringing in *too much* elasticity, which might send the wreck careening back into the path of oncoming racers.

**Styro Barrier at the Entrance to Pit Road:**
Sam Hornish Jr. lost control in turn four and spun off the racing surface in the direction of pit road. Even as his speed was decreasing, he hit the barrier that separates the racetrack (to the right) and pit road (to the left). Fortunately the wedge-shaped barrier was completely composed of crunchable foam bricks.

**Drivers' Earbuds:**
Drivers tend to adhere to silent rituals as they make final preparations to get into the racer. Rings are generally removed and handed off for safekeeping. Hair is swept back out of the way and earbuds are gently socketed into place, and the fireproof balaclava (or "head sock") is smoothed into place in order that the helmet fits comfortably.

The earbuds are essential for hearing the spotter, the "race strategist," and race control. A safety engineer has embedded tiny devices—a different one in each ear—that are capable of measuring violent g-force shocks that a wrecked driver's head absorbs. By "averaging" these accelerometer measurements, medical personnel can infer the degree of trauma that the interior of the brain (near the brain stem) has likely experienced. Drivers have been known to "wave off" impact injuries with an "I'm fine," only to later suffer from internal hemorrhaging. Now, an early reading might be cause for a driver to spend some time under observation until his well-being can be established.

When this tarp is deployed it is usually emblazoned with the Safety-Kleen logo. Safety-Kleen has a handy little outpost in the Gasoline Alley garage area where race teams may bring spent motor oil, transmission fluid, gear oil, and other "recyclables." The "diaper" is draped under a wrecked car so that as it's lifted onto the "hook" and carried away, it doesn't continue to dribble those same lubricants onto the track.

The damaged Red Bull car is hoisted in the "spider sling."

This wrecked racer shows where the tether system allows the broken wheel and suspension assembly to droop safely with the rest of the car. Fluids are being collected in the "diaper."

**Recovery Kit:**
Here at Nazareth Speedway, "recovery experts" are schooled in the specifics of IndyCar handling. A few hooks, a "drift pin," and some heavy Dacron line can be secured to various attachment points on the wrecked car to be carried away dangling in the "spider sling."

The safety workers from Indy are well aware of the way to clear a damaged (but still precious) IndyCar without further damaging the car. They supervise the training for IndyCar races at other tracks.

# Q&A WITH BILL SIMPSON
## SAFETY GURU

*Bill Simpson is the seminal designer and fabricator of modern racing safety equipment. He sold his original company, which still bears his name. After the specified "non-compete" period expired he began a similar safety company—Impact Racing. He subsequently sold that business as well.*

*He has hundreds of testimonials from drivers and crew who were protected by products that he and his companies have developed: helmets, harnesses, gloves, boots and coveralls. Drivers trust him.*

*He once (famously) walked a group of skeptics down to the grassy area at turn one at Indy, doused himself with gasoline, and lit himself afire—all to demonstrate the effectiveness of his new, patented racing suit.*

*The title of his hilarious biography perfectly defines him:* Racing Safely, Living Dangerously: The Hard Life and Fast Times of a Motorsports Mogul. *His co-author, "Bones" Bourcier, is pictured with him on page 70.*

'm an eighth-grade dropout, but I was a racer for a lot of years. I quit driving race cars in 1977. I competed here at the Indianapolis Motor Speedway. I qualified eighteenth and finished twelfth. That's when they had a hundred cars lined up to qualify. I invented the parachute for a dragster in 1958. My company just evolved. As the need arose for something, we just made it. There was nobody using parachutes; they used *nothing*. In those days they'd use seat belts out of passenger cars.

## ON IMPROVED DRIVER SAFETY:

There's not one thing you can pick out and say "that's responsible for driver safety."

First of all, the cars are a lot safer than they used to be. Secondly, the drivers' personal safety equipment has made monumental leaps in the past twenty years.

Racetrack facilities are safer. There are a myriad of things. You don't put your finger on one thing and say *that's* the reason guys are not getting hurt as much as they used to.

## ON HIS MANAGEMENT STYLE:

Ask anybody. I'm a dick. But I make good stuff. Everybody knows I make good stuff.

You ask my employees . . . I mess with them until hell freezes over, because they don't do it "good enough" . . . Ever! I'm anal, man.

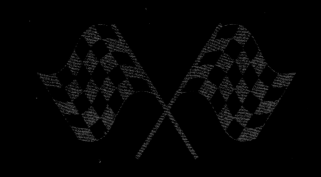

# 7
# THE RACE TEAM

**THE SETUP** for race day is a coordinated effort by the dozen or more key specialists who are under the direction of the team leader or head engineer. These specialists busy themselves in the confines of their designated pit area and may (or may not) be among the mechanics, fuelers, and tire changers that are fire-suited and helmeted so that they may go "over the wall" to service the needs of car and driver in the heat of the race. Prior to the race they will attempt to optimize (tweak) hundreds of variables to meet the track and atmospheric conditions at hand. Some of these "setup" factors include transmission gear ratios, aero package, dampers (shock absorbers), binders (brakes), tire composition, tire pressure, pit strategy, and fuel strategy. Competitive factors leading up to race day can include wind-tunnel testing, computer simulations, tire testing, driver conditioning, pit crew fitness, and hours of practice in hope that the driver can observe the exact speed limit on pit lane, bring the car to a perfect stop in the pit box, have all four tires changed, and fill the tank with up to 18.5 gallons of ethanol—all in about seven seconds! That accomplished, the team hopes for ideal "launch control." The car is lowered from its jacked-up position, the crew clears away old tires, tools, air hoses (and themselves), fueling hoses are disengaged and cleared (a crewman spritzes a precautionary jet of water on the fuel-fill area and on a set signal), the driver *perfectly* engages gears, revs, and then accelerates as he (or she) spins the new tires (it helps to warm them) and aggressively joins the line of traffic leaving pit lane.

Panther Racing preparing to do what they've rehearsed a thousand times.

The "dead man" is using gravity after a fueling to drain the "dregs" left in the hose back into the supply tank.

A fist pump from Mike Griffin of Panther Racing after a smooth tire change.

Showing all the parts of a practice pit stop/tire change.

This crewman, whose main job "over the wall" is tire changer, has used his welding skills to fashion a custom "nose wrench" for use when that assembly must be renewed.

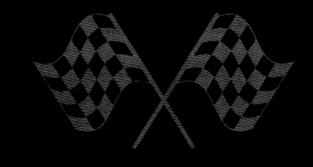

# 8
# DRIVERS

**RACING CHAMPIONS** frequently stay involved with the sport after hanging up their racing shoes. Mentoring the next crop of drivers is something that adds zest to the sport. Not every past winner has the eye to identify and the disposition to develop the next top driver, but many do it quite successfully. In some cases, these "teachers" might serve as a team owner (such as A. J. Foyt) or as a "driver development" coach for individuals, race teams, or the Indy Racing League. Young drivers can now "learn the trade" from past Indy 500 winners with names like Mario Andretti, A. J. Foyt, Al Unser, Bobby Rahal, Parnelli Jones, Danny Sullivan, Dario Franchitti, Gil De Ferran, Eddie Cheever, Johnny Rutherford, Rick Mears, or Arie Luyendyk.

The driver's world during a race can be a lonely, solo effort—much like championship golf, tennis, or boxing. Pressure to succeed can be intense, knowing that the world is watching individual performance "in real time." Having the guidance (frequently in the driver's earpiece) of one who has succeeded in similar circumstances can be a valuable tool.

Clearly, racers are a breed apart. They speak of the joy they experience when they match their driving mastery and their well-balanced machine against like-minded competitors. The rest of us are puzzled and amazed at their passion and skill.

They've won with go-carts, formula racers, motorcycles, sprint cars—on ovals, road courses, and hill climbs—and have a wall full of sparkling trophies to show for it.

Juan Pablo Montoya, winner of the 2015 Indy 500.

These small computer "thumb storage drives" are labeled with individual drivers' data. They were located in proximity to safety and "accident recorder" equipment, so it's thought they may hold medical information such as blood type and drug allergy.

**Rookie Stripes:**
The three vertical stripes here on the "impact attenuator" (crush box) mounted behind the transmission give drivers who are overtaking a slower vehicle a heads up: Caution! This is his first year out here!

# Q&A WITH LYN ST. JAMES VETERAN INDY DRIVER, ON TRAINING AND FITNESS

*I spoke with the 1992 Rookie of the Year on Race Day, 2014.*

## Q: WHAT CAN DRIVERS DO TO PREPARE FOR RACING AT THIS LEVEL OF COMPETITION?

It's really both physical and mental. You really divide them.

When you're training, the "physical" is more about endurance. So it's "enduring strength," it's "enduring cardiovascular."

The way I explain what it's like to drive a race car, like an IndyCar:

Put on a warm-up suit and get on an exercise bike. You should have a helmet on. No iPod, no music. You're not looking for something to relax you or distract you. Take two ten-pound weights and put them in front of you. Start pedaling and get your heart rate up to about 85 percent of your "max." That's what we maintain the entire time we're in the car. They've actually fitted heart monitors on hundreds of race car drivers to confirm this.

Taking your two ten-pound weights, look straight ahead and focus on something that you put in your line of sight. Turn those weights as if you're steering a race car. If it's a road course, you go both right and left. You have nothing to rest your arms on—no elbow rests or anything. You do that while someone with a ball-peen hammer is beating lightly on your body, simulating the vibration and energy you are absorbing in the car. This is what's happening all throughout your body; especially through the legs and torso and even through your arms, you're feeling that while you're holding the wheel.

You do that for forty-five minutes or so. That's kind of "the experience." As you're doing all of that, you have to make . . . split second decisions. It's a bit hard to simulate that, but you can do it.

So how do you prepare yourself to do that without draining yourself physically and mentally? You do that by those forty-five-minute or one-hour cardiovascular routines. You also have to lift weights. You look for the maximum you can lift for an extended period of time. It's an "endurance" thing rather than your muscle-bound stuff. Physically you have to train so that your body is prepared for the environment you're going to be in. And mentally—that's the harder part—to concentrate and to train yourself mentally, to not get physically drained when you have to concentrate for long periods of time.

I became aware back in 1988 or 1989 of a group at McGill University that had created a training protocol; included in that system they eventually developed a software program called Mind Shape. It was about long-term memory and short-term memory. It's not a "game," but it's kind of like a game in the sense that you have different curricula . . . on the computer screen. You keep track of your scores. You can store your performances in a database. If you wish, you can compete with some of the other people. You don't know who they are, so in that sense it's a little like a game if you want to use it that way. There were different settings for short-term memory or long-term. In one exercise they would flash five numbers on the screen and then they'd go away. Then they would ask: could you remember those? It really helps in exercising our mind and our brain.

They also had another deal—a big "light board." Jim Leo has one at PitFit here in Indianapolis. You stand there and focus on the red light in the middle. The lights flash, and you have to hit the exact one that's lit. You time this. It's "reaction time" and "anticipatory reaction time" that you're measuring. There, visually, you can exercise your eyes. You're looking peripherally. You don't really turn your head, because you don't turn your head much in the car.

You're training both your mind and your eyes. It's really the most important thing to have, for top performance when you're at 235 or 240 going down the straightaway getting ready to go into the corner at 220. Physically and mentally, it's draining and demanding.

More muscle doesn't make you go fast. You don't "muscle" the car around. You have to be very precise.

## Q: ARE THE DRIVERS ON THE STARTING GRID TODAY EXCEPTIONAL "1 IN 100" ATHLETES?

Absolutely!

I think so. If you look at them, they *are*. Look at the leanness. Leanness is critically important. If you have too much fat around your muscles then you "overheat" sooner. That's the biggest problem with that. Fat "insulates" muscle. What you want to do is be able to dissipate heat. You can do that when you develop a leaner body.

The training and conditioning required is really hard. I was never a great athlete. I can't run really well. I'm not physically the kind of athlete that's going to play baseball or football. However, there are "athletes in their minds" as well as "athletic bodies." It's a different execution of the word "athlete."

# 9
# SPONSORS

**A TRULY** sad sight at Indy is a race car without a sponsor. Sure, there are small stickers and logos here and there on the car, but the choice real estate—the sidepods—looks like a vacant lot. It's a very telling sign. It signals that the team owners have probably been bankrolling their campaign with (lots of) private money. They've acquired a couple of cars, contracted with an engine supplier, found a speedy driver, assembled a crew, paid all the entry fees, and *finally*, in the weeks before the "big show," qualified for the Indianapolis 500.

Even at this late hour, even if slower qualifying times put the car in the back of the pack, sponsorship will be frantically sought. The faster, traditionally more successful teams already have sponsors that support their efforts from year to year, and their financial commitment is for the entire racing season—not just the Indy 500. It's almost a certainty that someone *will* contract with our unsponsored entrant. This, most likely, will be a one-race arrangement with options for a deeper, continuing relationship if the car finishes well. Signage, decals, stickers, caps, helmets, shirts, jackets, and race wear, all in glorious color schemes, can materialize overnight when a sponsor is brought on board. Much lively brokering, agenting, and negotiating occurs in the precious few days before the race.

Two-time Indy winner Arie Luyendyk told us of his two winning race cars that are now showcased in the permanent collection of the Speedway museum. The proud sponsors were fellow Dutchmen who owned a meatpacking operation in Wisconsin. The record books (and the cars themselves) will indicate (forevermore) that he twice crossed the finish line in a Provimi Veal Company Special. This "jackpot mentality" has existed for a century at Indy. A company—or a group of entrepreneurs—will often take their entire promotion, marketing, and advertising budget and "roll the dice" with an Indy team. Visibility and bragging rights are what they seek (not to mention the adventure!). Racing immortality is theirs if the car should win.

It's perhaps unfair to defame the red car as "sadly without a sponsor." It could well have been in the process of being repaired, rebuilt, or reassembled. The fact remains that the quest for sponsors and "associate" sponsors often continues right up until the morning of the race.

In the mid 2000s, the Speedway brought the entire field of drivers who had qualified for the race to New York City for "media day." A gathering was staged in Times Square and featured honorary "marshall" for the race, General Colin Powell, the drivers in full race gear, and even Tony Kanaan's 7-11 "Big Gulp" race car. Such showmanship promoted the televised race and made the drivers available at a nearby venue for individual interviews with the world press.

Corporations, products, services, and causes of all types have chosen Indy to help spread their message. Obviously, the Indy "ruboff" for automotive wares is a comfortable fit for many.

Tony Kanaan and Scott Dixon—a winning pair of teammates at Target / Ganassi Racing.

**Target / Team Ganassi:** This durable partnership seems to be a textbook case of "branding."  Driver Scott Dixon exudes the winning spirit. Tony Kanaan is an equally charismatic figure.  Owner Chip Ganassi obviously enjoys every aspect of the sport.

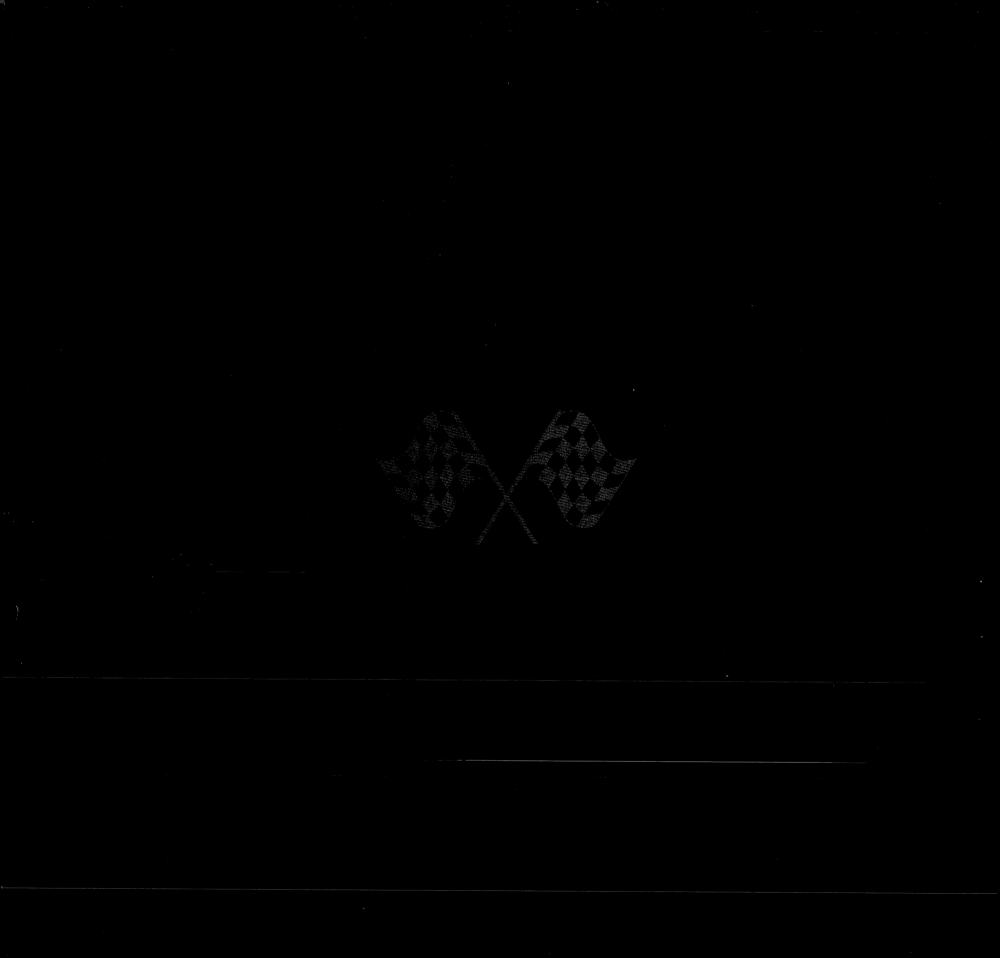

# 10
# RODGER WARD (1921–2004)

*Won Indy 1959, 1962*
*Date of interview: February 2004*

### RESPONDING TO HIS EARLY REPUTATION AS A "HARD LUCK" DRIVER:

That's true! In 1958 I finally "got my act together," and we ran pretty well. We ran well in '59, too, so it was a good situation for me. I think I realized that I had to get my act together in terms of driving a racer. I did that and ran pretty good. Back then we had the bricks—still—on the front straightaway; the rest of the track surface was good, but that front straight was kind of a hectic place because it was rough. You had to drive it very cautiously. The rest of the track was kept in first-rate order—it was really smooth and great. You had to drive the roughness of that front straight, but it was fun.

### THE MAN HE LEAST LIKED TO SEE IN HIS REARVIEW MIRROR:

Jim Rathmann! [*laughs*] He was the toughest guy I had to race. We were good friends. He was a good driver—a great driver—and we had a lot of fun together. Tempers could flare between drivers occasionally. It happened, but you tried to control yourself, make sure you didn't get out in left field.

### COMMENTING ON HIS CHIEF MECHANIC, A. J. WATSON:

Ah, he was absolutely super. He took care of the car and everything else. And it was fun to drive for him.

### FAVORITE INDY MOMENT:

That had to be the first year I won. That was really something. [Jim Rathmann finished second in that race.]

### HOW HE'D LIKE TO BE REMEMBERED:

That I was a good sport. I enjoyed what I was doing.

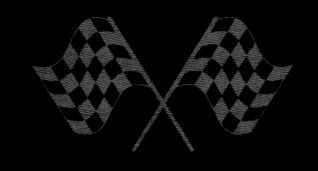

# PART II
# INTERVIEWS WITH THE WINNERS

Rodger *Ward*

**INDY WINNER
1959 - 1962**

**NATIONAL CHAMPION
1959 - 1962**

Chris Economaki had this to say about Rodger Ward: "Well, he was a very good friend and a very skilled driver. I did him a favor and got him a race car to drive in a Formula Libre race at Lime Rock (Connecticut, 1959)—a race that he won. He rated that victory on a par with his Indy 500 wins. Yeah, it was a midget against Ferraris, Maseratis, and other high-powered sports cars. Everybody laughed at him when he unloaded that morning—and later couldn't fathom the fact that a midget had beaten all those cars in the race that day."

The detail here (see previous page) reveals a good bit about Rodger Ward's era. The roadster, of course, has the engine in the front. One can only imagine the heat in that cramped "foot box." It has narrow grooved tires. The suspension has "up and down" shock absorbers—as opposed to the "fore and aft" used now to make the car sleeker and lower. He wears simple goggles and helmet. The coaming of the cockpit has two names of equal size—his and that of his chief mechanic A. J. Watson.

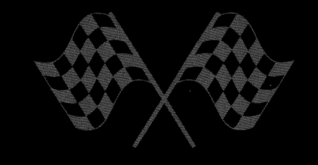

## 11
# JIM RATHMANN (1928–2011)

*Won Indy 1960*
*Date of interview: February 2004*

IN CALIFORNIA, when I first started racing, in 1946 or 1947, I guess it was, I wasn't old enough. I was sixteen years old, so my brother who is three years older than me loaned me his driver's license so I could get into Carroll Speedway. That was my first race. Still, I had to fight to get in there. I looked too young. I had long blond hair—well, not long, but I had *hair* anyway (all over my head). We got into a fight when they wouldn't let me race. Finally, they let me race. I just left my name that way—from then on I was Jim and he became Dick. I own some Chevrolet stores as Jim Rathmann, but my legal name is Richard on all the documents.

I was racing in Los Angeles; I was leading in the championship out there but there was no money. I saw an ad for a race at Soldier Field in Chicago with a $10,000 purse. I jumped in my truck and hooked on the race car. I had a super car. Great car (I built it myself)—a hot rod. I got to Chicago and started racing seven or eight times a week all over the Midwest. I made $57,000 one year, and $63,000 the next—not bad for three or four months' work—it gets cold up there. I had a friend named Anthony that helped me—a kid I grew up with. My brother came along and eventually he started driving. That was a couple of years after I did. I won hundreds of races—stock car races, hot rod races, and all kinds of races.

I've done so much stuff other than race driving. I knew all the original astronauts, was in the treasure business, and I won the world championship in go karts. I've done a lot of things in life. When I was a little kid I played in freight train yards. I was about twelve to fifteen years old and traveled all over the place. At nineteen, racing at Soldier Field, I was making fifty or sixty thousand bucks a year. I had a pretty good life.

I'd always wanted to race at Indy so I bought a Maserati chassis and put an Offenhauser engine in it and went to Indianapolis in 1949. It had a bad rear end on it, so I jumped into another car and made the race in it. Pat Flaherty was going to drive that one, but he borrowed *my* car to go to Chicago to make some money. While he was gone I took *his* car. That's how I got in the race.

From then on I ran fourteen of them at Indy. At one point I got outlawed because I was running at "outlaw" tracks. I guess I was money-hungry in those days. I wasn't supposed to be running that many races. I was making good money, $1,500 a day sometimes, running short tracks. I was making a lot of money and others were sort of jealous of me. I was making more than the guys that were winning Indianapolis for a few years there. Anyway, I was sort of an aggressive kind of guy. I like money. I've always been aggressive in business. In '48 or '49 I went to Miami and had a big business down there. I had a Chevrolet deal in Melbourne, Florida, and expanded that. I was the largest Volkswagen dealer in the United States gray market. The cars were not legitimate. I just went to Europe and bought 'em myself, and then sold 'em. I was a Lexus dealer as well as a Honda dealer with my son. Now he's running them. I can't stop working, though; right now I'm sitting here at my desk.

I've owned cars. Gus Grissom, Gordon Cooper, and I owned Indianapolis cars. I could never get a driver. I was disgusted with the drivers we had because they'd never "push the pedal." I tease the astronauts about the welding on a spacecraft. Race cars have better welding. Every welder and race car builder is proud of his workmanship and how he lays a bead on a piece of metal. At NASA it looks like a bunch of bird crap on 'em. All of them, all the spacecraft, they look horrible. There's a guy named Hatch who builds race cars for Daytona; he's a real good welder.

In racing there was dirt, but it never really bothered you, [since] you've got goggles on. The only thing that bothered you was that nitromethane fuel. You'd cry for twenty minutes if you started in the back with all the fumes on you. That's why I used to fight so hard to sit up front. Once you get going with the air circulating around you it goes away.

It's a thrill! What were there? Two hundred fifty or three hundred thousand people in the grandstands? It's a thrill just to see that—everybody cheering and hollering and jumping around. But you don't pay attention to anything but the guy in front of you.

Racers say it takes balls to drive a race car. Of course you can just get in the car and go out there and stroke around with the rest of the field, but if you're motivated and you're a real racer . . . I started in thirty-second once and I was leading it in forty laps—all the way from the back. I've done that a couple times. If you want to race and be a "charger," that's just something that's in different people. You've got to stand on the gas. It's a thrill—like kicking a football or throwing a baseball or something.

You don't want to lose your cool in a race car. You'll do something stupid and get hurt. It's something you don't want to do. I've seen a lot of guys with tempers, but it doesn't get them anywhere.

I got mad, too, but I've got enough brains, you really just want to get around the guy. I don't know what it is. If you start fighting with a guy on a short track, you can spin him out and not hurt him. He might get a broken leg or an arm, but nothing serious. But if you're going *fast*, say at Daytona Beach and you want to smack him in the ass-end, you don't want to do it there 'cause you could hurt the guy or hurt yourself. Things happen fast when you're running close to two hundred miles an hour.

It's all aerodynamic now. The car's got to be right. I won both the IndyCar races when we raced at Daytona. In 1959 we ran 172 miles an hour [on] average. We had those skinny old tires then, and you couldn't go very fast around the corners. You don't want to get mad at somebody . . . especially in an open-wheel car. I broke my back in Milwaukee one time "just playing." I was racing a guy, and his engine blew and he dumped oil all over the track. I was too close to avoid it. I was "pushing him," just for fun.

Daytona was really different from Indianapolis. At Daytona we were running over 170 right off the bat; at Indy at that time we were running about 135. Daytona, of course, is real high banking. Indianapolis is flat.

You've got to be determined. Hard. You've got to tell yourself you're better than the next guy. Go straight to the front. Talk yourself into it—if you've got a halfway decent car. Determination.

It didn't matter—quarter-mile tracks, banked, flat, Soldier Field, Milwaukee, Rockford, Illinois, Anderson, Indiana, Indianapolis, some dirt tracks—all those places. It was all "inverted starts" where the fast car started in last place. You have to pass all those cars to get to the front and win the race and do it in twenty-five laps. That made some money and made a good show. Some of that was open wheel, but I also got into stock cars—jalopies—later on.

Open wheel is a little more dangerous because you've got to watch. You can't get as close as you can with fenders. You can body-slam a guy with stock cars: taxicabs! With open wheel if you get together . . . the wheels touch and it's like being shot out of a cannon. The wheels grab and you go up over the guy. It's no good.

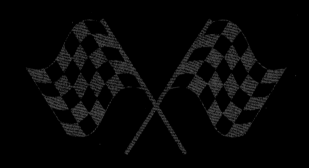

## 12

# DAN WHELDON (1978-2011)

*Won Indy 2005 and 2011*

DAN WAS *a very popular young Englishman. He tragically lost his life in a high-speed racing accident at the Las Vegas Motor Speedway in the last race of the 2011 IndyCar season. The man who led his quixotic journey from a geeky rookie to a two-time winner of the Indianapolis 500 had several qualities important in a driver, but one stood out to me most of all. This persistence (he bounced from team to team) was also reflected in his ferocious drive for the checkered flag.*

Arms raised in victory.

All his hard work has paid off. He wears the laurel wreath.

Concurrent with Dan's spectacular 2005 victory, another phenomenon was born—"Danicamania." For both drivers it represented a coming of age. Photographers (such as I) struggled to be in two places at once at the race's end: with the winner Dan Wheldon and among the celebrators as Danica climbed from her car. She was embraced by team owners David Letterman and Bobby Rahal. Dan Wheldon, ever the gentleman, waited patiently at the "Yard of Bricks" finish line with the huge laurel wreath around his shoulders until ESPN and the rest of the world press realized who actually won the race. Danica is a very good race car driver. Her stellar performance at the 2005 Indianapolis 500 (she came in sixth) earned her the Rookie of the Year award. She has a knack for avoiding trouble on the racetrack. In fact, she holds the record for completing fifty consecutive races (start to finish) without retiring for accident or equipment failure. I made a decision to extensively cover her pre-race activity on the starting grid before her 2005 Indy start. She was mobbed by media after her performance.

Danica, suited up and relaxed.

Autograph day.

That's a pre-race hug from team partner David Letterman.

Danica's hand grasp minutes before: "…Start your engines"

Danica's "game face" minutes before "drivers to your cars."

Dan on "autograph day" while with Andretti Green Racing after Michael Andretti's retirement from driving. He was not afraid to stand out in a crowd; he was an early adopter of the fake Mohawk or "faux hawk" hairstyle. He had not yet adopted his signature white-framed sunglasses. With his first Indy 500 win accomplished, Dan has achieved full "rock star" status.

Now, driving for Ganassi, he's let his hair grow and adopted those cool shades.

Danica racing for GoDaddy.com.

# 13
# RICK MEARS

*Won Indy 1979, 1984, 1988, 1991*
*Date of interview: March 2004*

I ALWAYS try to be a kind of "day in and day out" driver—not hot one day and cold the next. Consistency—we as a team strived for that. One of the stats that I like is that of the fourteen or fifteen years that we ran at Indy I started on the front row eleven times.

Qualifying there is as important as anything I've done. The race itself is a piece of cake—we've got five hundred miles to get it sorted out. Yet qualifying day—it's four laps—if you blow one corner you've blown the whole run. Now, if you are in contention for the front row— or the pole—you get your four laps in and put it in the "show" . . . and go on down the road.

But if you're "in the hunt," that's when the pressure starts. It's four laps at a racetrack that has no room for error. There's a fine, fine line if you're trying to run quicker than everybody. There's no room for error, so the pressure is great. The team I was with—the team I'm *still* with—they always had me in the hunt. The pressure was always on—it never let up.

There are two races—the first is qualifying and the second is the race itself. For me to satisfy my own ego, I guess, I want to sit on the pole—and secondly it paid pretty good. Mainly qualifying lets me know that I'm in the hunt. I'm gonna be competitive in the race. It's kind of a yardstick. I know if I qualify "way back" I've got a lot of work to do on race day. If you can sit on the pole, your sponsors— everybody—can utilize that for a WEEK. We've got a week between races. You win one race—it lasts for a week. So—you sit on the pole; it being *Indy*, and you're actually sitting on the pole. You probably get more press throughout the next week than winning the race would. It was good for sponsors and secondly it was good for me.

Qualifying is one of the ways of "paying the guys back" for the hard work that they're doing or gonna do. Any time I've run a lap and I saw the smiles on the guys' faces—that was kinda my payback. Any time I've run a good lap—whether it's Indy or anywhere else—if I've run a good lap and I saw the smiles on the guys' faces then that was *my* payback.

When I started—before it became as competitive as it is today—you could work all year long getting the dollars together, getting the equipment together. If you put a decent "package" together you could go out and "make the show"—run the race—and do that for *years*. It *is* a big thing just to run the race—period. You're one of thirty-three in the country—in the world—that made the show.

However, my first year there I didn't qualify. I didn't make the show. I was trying to qualify a four-year-old car—we were dollars short for updating equipment—I remember sitting on the wall thinking, *Maybe next year I'll have the experience . . . maybe I'll be back with a better team and more up-to-date equipment.* Little did I know at that time that the next year I'd be back with the best team in the business, and not only qualifying but qualifying on the front row.

Early on I was running "off-road" racing. I was also working construction—operating a backhoe. Rainy days there was no work—no money coming in, no paycheck. We were racing for fun on the weekends, maybe making a little prize money. It was always unbelievable the way it would work out. If you won a desert race and get "x" amount of dollars from that—and what "stickers" you had on your car—you got money from those. And that would all come in at different times; the checks would just kind of show up in the mail. It never failed, the timing was always perfect. You hadn't worked construction for a week, it had been raining, you couldn't work—bills were piling up. All of a sudden you open the mail—here's a check, one that you'd missed. It came in a month or so later—you talk about Christmas!

On a Colorado Motorcycle ride with Wally Dallenbach, Penske, Parnelli, Gurney—thirty or thirty-five guys from racing—Penske and I were parked pretty close to each other, and we got to talking. He said, "Would you ever think of driving for us? I'll tell you what . . . after this—why don't you give me a call?"

Just like that, out of the blue. I nearly fell over. So needless to say, I followed him very closely the next couple of days [*laughs*]. If he fell off I helped pick him up—dusted him off, straightened his handlebars,

whatever it took [*laughs*]. So we met at Michigan not too long after that—at the Michigan race. We sat down [and] he said, "This is what I have in mind . . . Mario's going to be chasing the World's Championship with Lotus—so this is your race—I want you to stand in for him. I'll guarantee you at least three 500 mile races—and when Mario 'lands,' you'll drive the third car."

Needless to say, I didn't have to think about that very long.

I've been with Penske since '78. That year we ended up running (instead of the six races that had been promised me), we ended up running ten races—('cause Mario's schedule was getting complex; I guess we were doing okay, too)—winning three races. The first was a flat track: Milwaukee. The second was a road course—Brands Hatch in England—and a high-banked speedway in Atlanta. So I had "covered the bases" as far as tracks. The rest is history. The next year I was "full-time."

It wasn't a lot of money. But for me, I couldn't have asked for more—it was good. But look—before that I was making nothing.

Indy was always one of my favorite places. It's obviously been very good to us.

Especially qualifying—it's so technical. It wasn't the "raciest" of racetracks—because of the narrow groove, the narrow pattern, [it was] difficult to pass. Cars come into play more—but [it's] very challenging. The first thing that jumped out at me the first year I went to Indy, I thought, *Whoa! This is different from all the rest.*

I didn't grow up dreaming of going to Indy, like some families [do]. That's a tradition for many. When I did go to Indy, I felt out of my league. I had to learn what Indy is all about.

When I won at Indy it was only my second time there. I'd won some races before that at other tracks. As time went on I'd look around and start realizing what it takes to win at that track. You start looking around and seeing how many good guys have been there through the years and have never won it. And you think, *I've won this thing . . . and I want to win it again.* You go there four, five, six more years—and you don't win it.

By the time I won it a second time, by then I'd learned more about what Indy meant—to [one's] career—across the board. It puts you in another category ("two-time winners"), but the odds are getting worse of you ever winning it again. So that just kept on building. You get to Indy and run another three, four, five years, and you realize it might never happen again. You look at Mario [Andretti]—in there year after year after year.

So then I won the third—it's unheard of! It just got progressively bigger and bigger and bigger . . . and better. And then, beyond that—FOUR WINS! That's a small group—me, Big Al, and Foyt.

# MEARS DISCUSSES THE ACCIDENT THAT ALMOST COST HIM HIS LEGS

They hurt every day—but they're there.

Racing is what we love doing. I was anxious to get back into the car. I did, five months later. It was great therapy. It gave me something to think about besides my feet. I started thinking about the car and the team . . . and about what we needed to do.

If I didn't know what caused it, I'd be a little leery—but I know what caused it!

What caused it was a combination of things. For one, the track was what we'd call a "bull ring." We'd had a test there—we were just coming off my second win at Indy and we were in the hunt for the championship that year. We'd never raced there before and the testing we'd done gave us an advantage.

That little bull ring—it's hard to get a clear lap there. I came up behind [Bobby] Rahal and Scott Brayton. I could see ahead of me and if I passed them—there was a big enough gap—I had a pretty good run. I moved up to Rahal on the front straightaway—as I went by him we were three abreast. I had to key off Rahal's left-rear with my right-front, and peripherally I saw Corrado Fabi pull away so I thought, *We're clear.* All of a sudden I felt the thing hit—my left-rear on his right-front—turned me left into the guard rail. What I didn't know at the time—I found out later—sometimes when guys are warming up they'll stand on the gas. Just as Fabi went out of my view he stepped on the throttle, which picked his speed up and kept him with us. That threw the timing off . . . which wasn't his fault, you know? He had nothing to do with it. He just happened to be in that spot at that time. Again, looking back, hindsight—I shouldn't have put myself in that situation. Over-eagerness put me out of control.

Damage? It just crushed both feet. The guard rail came through the front of the car. The guard rail was the problem. If it had been a concrete wall—ah—it may have done a little damage but nothing I couldn't handle [laughs]. The guard rails were not set in good solid ground—it was a new track; the guard rails were kind of set in gravel. There was too much of a gap underneath. Part of the nose got under the rail [and] the post started to come up. Part of the guard rail came into the car and caught the back of the pedals and just wiped my feet back. And then, the car was still going sideways down the track, and with a wedge like a pair of scissors it caught the next post. The car spun along a little bit. The guard rail tore off the front of the car and tried to take my feet back in the other direction with it. I had the "toe strap" around me—used in case the throttle sticks. It tore the pedals off. The pedals ended up in pit lane; when it tore the pedals off that toe strap was wrapped around my left foot. I tore off both Achilles tendons.

## AT THE CANADIAN EMERGENCY ROOM. . .

I didn't know it at the time, but thank God that Roger Penske was standing there.

The doctors walked up and said: "We can't do anything" and he said, "Hold on a minute! Time-out!"

They got me out of there—I was in bits and pieces. They got me to the [US/Canada] border—a helicopter ride partway, to get me to Penske's plane. They got me to Penske's plane and took me to Indy, and the hospital team there put me back together.

Today, I can look back on my little *deal* there and see how it really helped the industry. It was after that they really took a closer look at . . . move the driver, move the seat back, added to the bulkhead, made a bigger crushable structure near your feet. Really, it moved the industry forward as far as safety.

"The Captain," Roger Penske.

## ON MARIO ANDRETTI'S REPUTATION AS TOUGH TO PASS:

If you go into the corner deeper than any man alive, you're going to be tough to pass. That was one of Mario's strong suits. He could get a car into the corner as deep as anybody. You take it to the limit. It's hard for somebody to go *beyond* the limit and still pass you. He was very good at that—and he wouldn't cut you slack. He's not going to leave the door half-open for you. You've got to create that opening. But he's fair. He won't cut you slack—but he won't take you into the grass. He's just a die-hard competitor. He's all business when he's in that car. That's what makes him so tough to pass. But that also makes it more fun when you get past him! You get a better feeling out of it. You always get more enjoyment out of something that's tough to do.

## ON THE GREATNESS OF HIS BOSS, ROGER PENSKE:

To me, it's several things—it's no one particular thing. First of all, it's attention to detail. It's the effort, the dedication. One of Roger's sayings is "effort equals results." He understands the team concept and working together. There's obviously a pecking order, but everybody's always "in this together . . . " The team strives to be better.

That's what his business is all about—improving. If you stay the same, you're going backward.

He doesn't get the credit he deserves—for what he's done for the industry and the sport. The man just never slows down. He's just an incredible human being. If people knew half of what he does—or has done over the years. He takes care of his people. He has very little turnover of people. He just instills that in all of his businesses. He's a great motivator. He has good people.

He knows *people*. He's great at surrounding himself with the right people. If they're good people they're going to be good at whatever they decide to do. So if he has somebody in the racing program, [and] they want to get into something else [and] they're ready to get out of racing, he'll say, "Okay, what do you want to do? You want to get into my car business? The truck leasing business?"

Say they want to get into the service end of the automotive dealerships—or something like that. If they want to, he'll put 'em through school, get 'em schooling to point 'em in that way.

## 14
# DANNY SULLIVAN

*Won Indy 1985*
*Date of interview: February 2004*

### Q: ARE THERE OBVIOUS POINTS OF COMPARISON BETWEEN DRIVING AT INDY VERSUS FORMULA ONE?

I couldn't even tell you what an IRL engine turns out in terms of horsepower—that was after my time. In its heyday the Formula One engine was putting out about 950 hp. The IRL, I think, was putting out about 900. A slight difference. You know that would've been slowed down anyway. As time goes on they keep reducing horsepower, otherwise you make the cars too fast. Don't forget the tire guys are working to make a better tire, and the shock guys are working to make a better shock; engineers are always working on better aerodynamics. If you didn't alter that, the cars would be going 260 or 270 miles an hour now. A little bit like the development of golf club equipment—eventually it would make the race tracks obsolete.

On the subject of car speed, when you neutralize the regulations, good drivers like to have a car that's a little bit more difficult to drive because it separates the good guys from the (sort of) "also rans." I'm not saying that to question anybody's ability, but if everybody can run around the track wide open, then I hate to say it, but anybody can do

it. Now when the cars were a little more difficult to drive and we're sliding around . . . we could debate this forever. It's similar to what people complain about the NASCAR restrictor plate racing. They want to put it back into the drivers' hands more. That's a universal complaint, by the way. It's not aimed at the IRL or any other series. The rules makers are looking to make it as equal as possible because it makes better racing. So you're caught between a rock and a hard place.

To go back to personal experience, [I will mention] a guy who's a friend of mine named Pat Bedard. Although mainly a journalist, Pat was quite a good racer himself and attempted the Indy 500 on a couple of occasions. He qualified one year, I think in the top ten in a "March." Everybody, including myself, were like: "Oh, shit!" . . . not because of Pat, because he's a very likable guy and everything. The problem is [that] you go and run four laps or two laps for qualifying—a lot of people can get their car set up pretty good to do that. But when you get out there in a race situation, [it's different]. The track conditions are changing (the track's getting greasy), there's rubber going down, now you're running with thirty-three cars, people are moving and breaking at different places . . . everything starts happening! There is no "memory bank" for him to draw on from experience. Pat had a huge accident! This is just me, Danny Sullivan, my opinion, [but] I think Pat got caught in a situation where maybe the car moved or slid, there was a traffic situation or whatever, and he had no resources or experience to draw on of what to do. That doesn't mean . . . I could have crashed in the same situation, so it just means . . . you know when you're running around with guys going 220 miles an hour, you want them all to have a little bit of experience and know what they're doing. I echo what a lot of other experienced guys are saying: "Hey, I don't want to be running with a guy who can just hold it down on the floor, who's got big balls and no brains and no experience, and maybe put me in a situation where I'm now in danger."

We want to sort out the men from the boys a little bit here.

## THE SPIN

Early on in the race, before the first pit stop, my car was not handling well. But Derrick Walker, of Derrick Walker Racing, he pitted me "out of step." He took a gamble, because Mario—when I went back out—Mario was only two seconds from lapping me after that pit stop and I "caught him back" within about thirty laps. My car was now really good.

I say I "caught" Mario, but the race has gone on now for a hundred laps or so. There was a lot of turbulence out there on the car, and I couldn't read the pit board very well, and you tend to lose track of time when you're out there, you know? It's not like you're counting every lap. [So] I radioed in (in those days, radios were not as good as they are now. *Nothing* is as good as it is now): "How much time is there to go in the race?" Now, I'm running second, right behind Mario's gearbox; it *does* make a difference.

I *thought* that they said "twenty-eight laps" [*laughs*] I was only off by about fifty laps, but at the time when you get down toward the end . . . Mario, by the way, is among the toughest sons-of-bitches to pass at the *best* of times. But if you're coming down to the end—say, the last twenty laps of the Indianapolis 500, he's going to be a complete monster to get by. So my car was working better off of turns three and four than it was in one and two. Why? They're pretty much the same at both ends, but the reality could have been . . . there could have been a breeze blowing across, it doesn't matter. But I was quicker across three and four. So I was drafting Mario really close down the front straightaway; if you ever look at the videotape from a "head on" or a "back" shot, I mean he virtually had me up against the pit wall. So I got alongside of him and as we went into turn one (now the track's different, but in the old days . . . [*laughs*] I never thought I'd say "in the old days") . . . and rather than the little apron that they have now, that was where the warm-up lane was. You warmed up right next to the track. It was a little bit flat with a slight inclination to go up the banking and, of course, the stripes were painted along there.

When we got into turn one, Mario and I were "wheel to wheel." I'm on the inside; Mario just "turned," [and] he drove me right down to the grass. Of course, because of the angle and going into the corner I had a shorter route. I pulled in front of him—I was quicker than him—so as I pulled in front of him and came back up on to the track, the white lines, the slight inclination of the track . . . whatever it is, a combination of everything there, it just "tripped" the car just enough to make it unsettle the back end of the car. I tried to "correct" the car—meaning "turning into the oversteer—and the front end started to "bite." I knew that was a big no-no, because if the front end bites it's just going to turn sharp right and you're going to go nose-first into the wall and you're history. So I turned the wheel back; I thought I had it, but the back end just kept on moving and going 210-odd miles an hour. Eventually I just lost it. There was very little downforce on these cars in the first place because you are trying to go fast as you can on a straight line with as little downforce as possible. I spun, and I just went: "Oh, shit!" I was so mad, because I'd just gotten into the lead of the Indianapolis 500, and it was gone. I spun and then I instantly got on the brakes, like, "Ah, shit." I wasn't concerned from a safety point of view. I just knew I was going to hit the wall. Then the smoke cleared and I was facing the turn two suites! I just took my foot off the brake—the engine's dead now—and I thought *What gear?* I know in those days—they probably still do it now—we ran two top gears. It was a five speed—I think it may be six in these cars, I don't know—but I had three "speedup" gears and then two top gears. If I take the speedup gears and I'm going too fast (I have no idea how fast I'm going now) then that's no good; with the top gears, if I take too "long" a gear—it's like a handbrake on the rear end—I'll spin again. Actually I took fourth [gear] and it almost got away from me again, but it jump-started it and I took off! Of course the "yellow" is out, and I said to Derrick on the radio: "Derrick, Derrick, the yellow's for me, the yellow's for me, everything is okay, everything is okay, the car is fine, I didn't hit anything!"

He said that I was about as normal can be, but my voice was about three octaves higher.

I made a pit stop and went back out, got back in line. When they gave the green again it took me about twenty laps to catch Mario, and I passed him in *exactly* the same spot. But this time when we went into turn one I moved him up a little bit closer to the wall, and he *still* tried to come down on me. I would have done the same thing; I mean we're talking about the lead for the Indianapolis 500 and he's a tough, tough, tough, competitive guy. I would've done the same thing; I mean nobody wants to give up the lead easily.

Before I caught him the second time, there was a scarier moment. Howdy Holmes and Tom Sneva got "tangled" going into turn one. I had luckily anticipated it and saw it coming, so I started slowing down. Sneva spun, Holmes went into the wall as Sneva spun . . . normally when a car spins it either goes right or left, so you usually go right at the car, but Sneva went straight down the track. I missed him by—I couldn't even see it—I thought I was going to hit him with my nose. I passed Mario soon after that, when they cleaned that up and I got past him.

## THOSE TIRES

After the spin, the tires were obviously "flat-spotted," but you know it spun like a top. It was a very quick spin. It just went *whish*, but because there was no downforce on the car, they were flat spotted. But if the yellow hadn't come out? I could have kept going. It wouldn't have been comfortable, but I would have been okay until the next pit stop. It would have vibrated some, but I wouldn't have had to slow up a whole lot.

Goodyear and those guys, they take those things away pretty quickly. They just don't want, in case there's a problem or something . . . I never saw 'em again.

I'll tell you something. I've told the story many, many times, but if I watch the tape, I still get goose bumps. I've probably seen the tape three thousand times. Every interview I did for years after that, they show the tape. I know the outcome. I know I didn't hit anything, but it still gives me chills.

# 15
# MARIO ANDRETTI

*Won Indy 1969*
*Date of interview: October 2015*

*The three-time United States Driver of the Year discusses the first time he ever saw the Indy 500.*

**IT DOES** start with a dream. I came to the States when I was fifteen years of age. My dream was in place. I had no idea how I would pursue it.

Two years later, my brother Aldo and I, my twin brother, we got together with a couple of other buddies and we started building a car. Fast forward to 1958, when an uncle of ours locally here, after listening to Aldo and me express ourselves about racing and so forth, he took us to Indianapolis.

Wow!

That was my first experience seeing the place, and I was eighteen years old. The car that we were building was almost finished. We arrived there and stayed right there at 16th and Georgetown. It was like a trailer park. We slept in the car. We had good seats, pretty much at track level at turn four. As we watched the race, at one point I said, "I'm gonna be here someday." It was much like when I was in Italy watching the Italian Grand Prix at age fourteen and I said: "I'm going to be *here* someday."

## WHEN DID YOU START GOING FAST? DID YOU ALWAYS HAVE AN ATTRACTION TO SPEED?

We had a '57 Chevy and we would just tool around. We got up to 120 miles per hour—as fast as that thing would go. We got in trouble a few times. We would rule the neighborhood—I can tell you that!

But what the heck, it's not racing until you actually get on the racetrack. You're not really experiencing it. That started when I was nineteen.

My brother and I used to do everything together. Aldo and I both had the same dream. Unfortunately, for him, he was injured very badly at the end of the very first season in 1959. It was an invitational, the last race of the season. After the accident he took a sabbatical. A year later he came back and raced for ten more years.

In '69, the week after I won Indy, I had a race in Milwaukee and he was in Des Moines, Iowa, racing a sprint car. That was going to be his last race. He had another big accident and that ended his career.

Then his life changed. He was laid up for a year. He went on to raise a family, start a business, and live with his family in a normal way. He was supportive of us but never raced again.

Aldo now has a "third generation" driving as well. His son John, John Andretti, has started at Indy, drives NASCAR. He's a great driver. John's son Jarett drives sprint cars now. I have the third generation on my side with Marco, Michael's son. The family continues life in our sport. It's really the only thing we know.

The great Indy figure Andy Granatelli (R.I.P.) is in the hat. The jubilant sponsor of the #2 STP Oil Treatment Special once famously planted a congratulatory kiss on Andretti's cheek as he climbed from the racer. A photographer captured the moment and soon all the world would see it, long before the word "viral" found its current use.

Mario's grandson and Andretti Autosport racer Marco Andretti.

Mario's son (and Marco's father) Michael Andretti.

As a teenager, before he became the experienced Indy veteran he is now, Marco receives some trackside wisdom from grandfather Mario.

## ON HIS ARRIVAL TO INDIANAPOLIS AS A ROOKIE:

I got out of driving stock cars at the local level—got into midgets and sprint cars in 1963 and '64. When I got to the USAC level in sprint cars and I won some races, that's where I got picked out to try a Champ Car, by the Dean Van Lines folks. They had Clint Brawner—he's the one who gave Foyt his driver's test at Indy. Eddie Sachs and Jimmy Bryant were driving for that team. These guys were icons.

I was invited to run some Champ Car races in '64, after Indy. Actually, I was hired *before* Indy, but Clint Brawner didn't think I was ready, so he told me: "I'll have you in a car right after Indy"—in the "Hawk." That was it. I started and then ran the rest of that season. I qualified [for the 1965 season] as a rookie. That next year, my first race in a rear-engine car was exactly at Indy. I had never as much as sat in one before. Up to then I drove a roadster—really, one of the latest design roadsters—nice lines. I did well with that. Then, my debut with

the rear-engine car happened. I took to it like a duck to water. I was in my element immediately. The car arrived at the track [Indy] a week late. Others had already been practicing a whole week. In those days you could take the driver's test up until Wednesday of the second week of May, which was the week of qualifying. My car arrived exactly on that day. I was "on track" that Wednesday. I got my test and passed with flying colors. I finally stopped biting my fingernails.

Can you imagine . . . me, no car, and everybody else is running around learning what to expect. I'd never even been on the track before, not even in a pace car. The first time I got on the track was in a race car.

Well, it's one of those things—you just gotta do it. When you get on the dance floor, you gotta dance! Anyway, it felt really right at home—me being at Indy. I had that mindset: just stay positive, just stay positive that everything would work out.

Luckily, everything was right. I was certainly with the right team with Clint Brawner, who'd seen it all. Clint had an assistant, Jim McGee, now known as one of the all-time great crew chiefs at Indianapolis. It was a seasoned team. I had the best of all possible worlds going for me. You *cannot* do it alone.

In '69 when I won I had the same team. I had really good luck and good times at Indy. In '65 I sat in the second row and finished third. In '67 I was on the pole! I could have won two of the easiest races of my life—and both times the cars had mechanical failures. In '68 we had a new engine—a Ford engine. It was a mistake—the thing blew up on the first lap.

Don't forget that in '65 I won the National Championship, as well. I was the youngest driver to do that. In '66 I also won the championship. In '67 and '68 I finished second in the championship. I would have had five straight championships. In '69 I did win my third championship. I won that one in August; the last race of the season was in December, so I really had a great year.

Sixty-nine was the only race I finished at Indy since my rookie year. I felt very comfortable. Indy was the first place that I felt that I knew how to drive. I love the place. The thing that was to my advantage was the team. Dean Van Lines had a solid relationship with Firestone. They were the Firestone "team of choice." Firestone was also looking for some fresh input. You know, they had all of the seasoned drivers like Parnelli Jones and Rodger Ward—drivers like that doing a lot of the "development," but they felt that they needed some new fresh input. Sometimes "adventurers" have a way of doing things, explaining things. They wanted to have this on the team of their experienced drivers, and I happened to be the one. Because of that I was getting a lot of "testing." I was driving thousands of miles of testing. Indy was one of their places. I was breaking records testing; I knew every crack on the track and I still do today.

Indy, as symmetrical as you think the track is, is very fickle. Don't take anything for granted. Every corner is *absolutely* different. With speed, the faster you go the more the difference magnifies. The wind factor is *huge*. The wind is going to blow differently on every corner. If you don't think you need to compensate, well, think again. That's why there are two flags. When you go down the straightaway, up on top of the pylon there's a flag; you can see which way it's blowing, and compensate. You go down the back straightaway, it's right in front of you on the grandstand. In turn three there's another flag that you watch, that you better pay attention to. So the bottom line is when you're really "at the limit" at Indy, believe me you better know what you're doing and what to expect. Like I said: It's not as symmetrical as you think.

The banking is only nine degrees. It's relatively flat compared to some of the oval that we run. Because of that, obviously, the onus is on the driver a lot more. You have a tendency to "drift" more when you don't get the banking "help" in the corners.

Of course we now run "flat" all the way around the track. You're going 235 or 240 down the straightaway—that's the way you enter and go through the turns in *qualifying*, not in the race.

G-forces? In qualifying we've seen 5.2.

When I was racing, in Michigan, for instance, when I qualified at 234.7 average on a one-mile track, I had 6.2 g's measured.

Obviously, you brace yourself. You make sure that your seating arrangement and everything else is really, really tight. You must fit like a glove. You must be strapped in very, very tight. You have "side support." You can't turn the wheel; you just try to hang on. Your body is being slammed to the side, and it feels six times heavier due to the G-forces. That's what people don't understand. That's why it's so important to fit yourself really, really well in the car, in the seat. At that speed you also have to be able to *feel*. Get the feel of the car—which side is shimmying or giving up; you don't want to lose anything. At the same time you have to be comfortable enough that you can steer the damn thing.

The relationship between drivers and engineers is the same as patients and doctors. I'll give you an example: the driver is the patient. The patient goes to the doctor and says, "Doc, here's where my problem is. I hurt here, here, there." So the doctor goes right to the point.

If you go there to the doctor and say, "Doc, I feel like shit. I'm hurting all over. I don't know where it is." So what's the doctor going to do? He's going to put you through tests and . . . God knows. So when a driver comes in, the more detail he can explain [regarding] what the car is doing, the easier it is to go to the "fix."

If you come in and say, "You know what? This thing is a piece of crap. I can't make heads or tails of what is going on," what is the guy going to do? Where does he start fixing?

That's where teamwork makes a difference. Being able to get it right and getting results . . . and not.

You hear this all the time: "I really, really gel with my engineer (or crew chief); we really understand each other." That's what it's all about; that's where teamwork comes in. A lot of people don't understand how important teamwork is in our sport. You know, the driver is usually either the hero or the goat.

The best driver in the world cannot produce miracles if the equipment is not "capable." You cannot do what the equipment is not capable of doing—you'll kill yourself.

You need to have the team around you. In a race, the team that "services" you, they can make it or break it. Personally, I've grown lifetime friendships with mechanics and others that I've worked with over the years. We'd created a real camaraderie. They knew that I was

Mario Andretti was spotted as he walked from the Andretti Racing garage in Gasoline Alley, prior to being besieged by fans seeking an autograph. He's prepared, with a pen in hand.

out there bustin' my butt [100 percent]. They gave you ten-tenths as well. We inspired one another.

Win or lose, we do it together. *No one* ever had more appreciation for the team than I did.

## THE ANDRETTI DYNASTY OF RACERS?

Let me tell you this: This is the business we know. The only business we know. When the family is together we bore the hell out of the ladies. Christmas dinner? We talk about racing.

The bottom line, what I have found is, from my standpoint. I cannot teach young drivers to go fast. That has to come from inside. From your belly. The only suggestion I can make is help them to minimize mistakes. That's the only thing I can contribute.

If I had to do my career over again, I guarantee you there are a few things that I would do differently. Perhaps I can point out a few "shortcuts" away from potential mistakes.

Personally, I have no regrets, believe me. No regrets and I count my blessings every day. But if I had to do things all over, some of the mistakes I made, like being too "exuberant" the first lap and stuff like that—I wouldn't do those things again. I would be more patient.

You know, you learn from mistakes, for sure. That's where I feel I learned, because I made *plenty* of them [*laughs*].

I made every mistake ever made. As time goes on, that's what you strive to minimize, through experience. That's the contribution I feel that I could have for younger drivers. If they ask a question you give an honest answer. They have to ask the question. It's the same thing with my grandson. I don't volunteer unless they ask the question. Sometimes when you "volunteer" they take that as criticism. Not constructive criticism—they can take it as just criticism. I've learned that. Then they resist you.

So now I say, "Okay, I'm there if you need me. You want an opinion? You just want to pick my brain a little bit? I'm there a hundred percent for you."

*The author had lingering questions concerning the '57 Chevy that the Brothers Andretti shared as teens.*

*Mario kindly responded the following day:*
*Q: What color? A: Red with white top*
*Q: Body type? A: Bel Air 2-door hard top*
*Q: Engine? A: 283 Duntov cam, 4 barrel carburetor, stick shift*
*Q: Had the boys modified it in any way? A: Added glass pack muffler*

# 16
# BUDDY LAZIER

*Won Indy 1996*
*Date of interview: February 2004*

I STARTED in IndyCar racing when I was eighteen. My background is in Formula V and Formula Ford open-wheel sports car racing. I made the transition to Can-Am in the late eighties. [I] then moved over to CART; the IRL didn't exist. For the first six, seven, or eight years I was driving for teams using what were basically their "show cars." A lot of times I had just a voluntary crew. We didn't have the budget to pay them. Eventually, I won a championship in AIF, which sort of grew out of Can-Am. For Indy, at first, I really struggled. At age eighteen or twenty I had attempted to qualify in a car my father and I owned. It had been given to us by the Machinists Union. The first attempt was with that three-year-old race car. The second year was another three-year-old car.

I was on the bubble and we thought it was going to rain the next day. Before the rain came on Saturday we went out and attempted to qualify with an engine that had seven hundred miles on it. You have to race on the engine with which you qualified. (We had a brand-new, fresh, qualifying engine, mind you, that was ready for the next day.) We got the run in just before the rain came.

We took that Saturday qualifying time on the advice of a lot of people around us.

I was on the bubble but didn't make it. It cleared the next day and we got bumped. There are lots of lessons learned by rookies and young drivers.

A lot of my experience was gained the hard way, and I'm proud of that. My skill set, down the line, was built on that early experience.

I finally made it in '91 [at the age of twenty-four]. I was learning my craft. You know that old saying "You don't need a Stradivarius to learn to play the violin?" [*laughs*] Really, where I was coming from, I was on the very bottom rung of the ladder.

The Machinists Union had cars they didn't need anymore; they were sponsoring another IndyCar team. They didn't even need them for show cars. They were that old. We took that car and worked on it and that was the car I used to pass my rookie test. It was their way to help out a young driver. The first car was an '87 March. I had '86 March's, an '85 that Bobby Rahal at Tru-Sports Racing had. I basically started my career with used cars. I was trying to find a way to break into the series. It was very difficult for a driver to break in. You could count the American drivers on one hand. Now for young drivers, we have a feeder series like Indy Lights. It's just awesome what the Hulman family and Tony George has done. Plus USAC, the United States Auto Club, has sprint cars and midgets; the career path for a young driver is a lot different now than when I was starting out. We were basically pouring our own money into it, and we were limited in that. I'd go into the market where the race was going to be way ahead of time and see if I could line up local sponsors, $5,000 or $10,000 at a time. We had people that had the same level of passion that I did. They weren't professional fund-raisers (they had a business background), so they went out and did whatever they could. A voluntary crew and a three-year-old race car, that's how I got started. It was hard.

I have a memory of that first season—it really makes me smile. We were going to Laguna Seca with this three-year-old Machinists Union's car—this was late 1980s or early 1990s. Almost everybody back then had graduated from trailers to "transporters." Our voluntary crew on the way to our first morning of qualification was packed into one car. We were driving up this big hill to the track and in front of us was this big cloud of dust. It turned out that it was our race car being pulled by an old "dualie." The truck and the "tow-behind" trailer had also been given to us by the Machinists Union. The fact is it couldn't make it up the hill. We had to get *another* tow vehicle in order to

With Buddy, it's a family affair. His brother Jacques and his father Bob made multiple starts at Indy.

tow our tow vehicle up the hill. We went out for practice, and after a couple of laps the whole car caught fire. We got back to the pits and were trying to put out the fire. I think it was [team owner] Doug Shierson at that time who came over to my dad and said: "What are you trying to do? Just let it burn!" [*laughs*]

To break into the sport we had nothing but pure passion. My father and I knew that if I could get a race car that was equal to [that of] the other contestants I could show well and win races. One thing that never wavered was my capability. So the question was, "How do we get over this mountain and win races?"

A number of years later I can look back—I've won *many* races. I've won eight IndyCar or IRL races, I've won the Indianapolis 500, I've been runner up there twice (in 1998 and in 2000). To think where we started to where I am now, and all the struggles in between, it's been a wonderful experience. Here I am, I'm thirty-five years old, and I feel I have another ten years of my racing career. It's been very, very good for the last eight years, and now here I am looking for opportunities. Certainly one thing a driver can do when he doesn't have a ride is to draw on this past experience. I feel confident that I can get over this part of my career as well.

## ON THE YEAR HE WON AT INDY:

I had a couple of things coming together. I'd driven for John Menard the year before. I had qualified for the 500 the first day. I was the third driver that year. We had a great race and I actually felt that I could have won. That was 1995; my car lost fuel pressure, and we had to stop early, but I felt we had a strong race car. As a race-car driver who's won races, I know what it's like to finish last. In 1991, my rookie year at Indy, Gary Bettenhausen spun out in front of me and I made an evasive move to miss him. I was out on the first lap. We had knocked the nose off the car and we didn't have any spares.

I learned exactly what it feels like to finish "absolutely last" at the Indianapolis 500.

I also know what it feels like to *win* the Indianapolis 500.

I know what it feels like to finish second, [as I did] in 1998. (I was lucky to finish that high.)

In 2000 I learned what it's like to finish second and feel that we should've won. We had the fastest car that day. We had seven pit stops, and the person that won had five pit stops.

I think I've experienced most of the emotions of the Indianapolis 500. It's always a roller coaster of emotion. It's just the most amazing racing facility. There's just so much that goes into the Indianapolis Motor Speedway—the history behind it that will always make it the most magnificent racetrack in all the world. Being a competitor and going through that roller coaster [of] emotions. You know they call the month of May the "year of May" because so much happens in those thirty-one days. As a competitor looking back on ten, eleven, twelve of those months of May, that's a tremendous storehouse of emotions.

## BACK TO THE 1996 WINNING YEAR:

In 1995 I had been racing with Menard. It seemed like I was slated to race with them in 1996. Then it was announced that the IRL had been formed. I was called to do a tire test: Firestone at Orlando with Hemelgarn Racing. We set all kinds of track records there. Then we went to Phoenix and set more track records. We were timed as the fastest that anyone had gone around the one-mile oval. We just had amazing preseason tire tests. Menard was still putting their plans together when Hemelgarn made a commitment to me. We had done so well in all these test sessions, so I went racing with them. We went to the first IRL race—the Orlando race—and sat on the pole. However, we had a suspension failure while we were winning the race. Then in the Phoenix race (still in the first year of the IRL-race weekend) in practice, I had a rear-wing failure. Coming down the straightaway, the rear-wing plate delaminated. This caused the wing itself (which creates all the downforce) to fail. I entered turn one not knowing that anything had gone wrong [and] did a huge pirouette and hit the wall a ton. I very badly broke my back . . . in multiple places, [including] my lower vertebrae and my sacrum. My sacrum had about twenty-five fractures, "spiderweb" cracks all over it. So I was very badly injured. This was eight weeks before I won the Indianapolis 500. You know, I was airlifted out of there. We know it was 99.6 g's. Nowadays we see a lot greater. The Indy Racing League has done so much to improve safety of these cars that now, 99.6 g's is an easy accident to walk away from. There was a lot of damage done. My recuperation started in Phoenix, where I was in intensive care for a week or two; then the league in their medical program had me air vac'ed back to Vail, Colorado, where I wanted to recover. I recovered in Vail for several weeks, and I think it was five weeks after the accident that we went to Indy to think about qualifying for the May race. Obviously the dominoes would all have to be in place, but we were thinking *One thing at a time. Let's see about qualifying.*

I had to be able to get on crutches before we had to leave for practice and time trials. Five days before leaving for Indy I had to be able to walk with a cane. When I went to Indy the bones had to have healed well enough so that the medical staff could say that you're no longer at risk to further injure yourself, that you're safe to race. I was healed just

They had just repaved the race track for the month of May, so it was as fast as I'd ever seen it. We had incredible grip and that was the last year of the turbocharged engines. We were just screaming. We were cutting off laps—wide open—in the 238 or 239 miles per hour. I think it hit 240 in the draft. We were just haulin'. It was a very, very fast race. You have yellows—cautions—that pull down the average speed of the race so I don't know if the record supports it, but it was a very fast race. For the first time in my life I had [a] car that could win the Indianapolis 500. With all the struggling I'd done I wasn't going to let an injury like this keep me from winning the Indy 500.

The only thing that's on your mind is what's in front of you. For me it was looking at each stage of the race. Okay, that was a clean pit stop, picking off race cars and moving my way through the field, and of course "staying clean"—making it to the end of the race. And when I got to the end of the race, I was able to let it all hang out. I made a couple of passes at the end of the race for the win.

You never know if it can happen again. For the last eight years I've been trying so hard to get back into Winner's Circle—especially in 1997, when we came so close, missing it by three or four seconds and in 2000 missing again by two or three seconds. When you take second place at the Indy 500, if you'd never won it before, it would be the greatest thing.

The Indianapolis 500 is all about dreams. In life there are so many dreams that don't come true. I wouldn't say "crushed" dreams. There are so many people that have gone there [and] given it all they've got, but they just didn't have the race car to qualify. That driver's life has now been "put on hold." He has an eleven-month quest to get back with a bigger gun, a bigger race car that would be capable of qualifying. Then there are guys like me that have struggled, then found a race car that will qualify, and go on to win the race. I've had dreams that have come true.

in time. The x-rays showed that. Fortunately in 1996 I was in the best shape of my life. I still keep myself in good shape. I was still hurting. I was still recovering from a pretty substantial injury. I'd hit the wall backwards, and at the time the cars were just not designed . . . did not have the diffusers on the gearbox that the Indy Racing League mandates. We didn't have a lot of these "crushable zones" then. In safety, every year there have been monumental gains. We started the race in fifth place. We [then] took the lead and won the race! What was so great for me was to go from the injury and recovery (I had [also] gotten engaged on May 1 to my girlfriend), and went on to win the Indy 500.

My goal in that race was to keep the race leaders in sight. We had a very fast race car, but it's not always the fastest cars that win. You have to be in the race at the end to win. I bided my time, I led at certain parts of the race, at other times in the race I was in the top five. Until . . . (I think Bobby Unser calls it the "blood and guts" of the race) . . . the last stint. The last pit stop.

Out!

You really let the car "all out."

## BUDDY COMMENTS (VERY SPECIFICALLY) ON THE DRAFTING EFFECT:

There is an effect, and it depends on whether the car that has pulled up behind has put himself "right on his gearbox" or if he is cocked out to the side. It's a very complicated aerodynamic, and every year it changes a bit. In open-wheel racing the tires and the rims are spinning at enormous rpm. It's huge. The wind reacts with that; it creates an enormous pocket of air. If the guy that's passing is working with the guy in front of him, he has the ability to either slow him down or speed him up a little bit. So if you're working with a guy, the two of you can go faster *together* than either of you could on your own. It's not as much a factor as it is in stock car racing. Our cars are much more aerodynamically efficient.

If two cars are cooperating, the second one might be at 70 percent throttle, or even "rubbing" the brakes a little bit (left foot braking), but generally it will depend on various factors. The first is *handling* (how, specifically, is each car behaving?). It could be that the guy out front is a little *loose* and the guy that's trailing is *pushing*. So the guy out front has to run a little higher into the entrance of the corner. The second guy would run a little lower through the entrance of the corner; you'd see them separate a bit and then get back together as fast as they can. The fastest race cars run today right in each other's wheel tracks, as close as you can get without making contact. A couple of feet apart right in the wheel tracks is the fastest those two race cars can go.

Along with that, the car that's trailing does not have as much downforce over its wing and through its tunnels. In the turns that second car has almost no downforce on his front wing; that's another reason you see them separate as they enter a turn. Sometimes the second car will drop his inside tire so the left side of the car is running in fresh air. Alternately they'll move the car to the right so the right-hand "third" of the wing is in fresh air.

At Indy, basically a flat racecourse, you'll see accidents because the cars are too close, and they're in that area where they have no downforce. They might enter the corner at the wrong angle and find that they can't recover.

They can't get grip in the car fast enough, so they hit the wall. It's very, very sensitive. The car that's behind has to make sure he's grabbing enough air so he can keep it stuck to the racetrack. That's one of the reasons you'll see them all kind of sidestep out to one side or the other to get just enough clean air to get enough grip to make it through the corner.

The car in front is dictating their line. The car behind has a "seat-of-the-pants" feel for the race car. He senses its grip and balance.

When you're "tucking out" a little bit, generally speaking you're also watching your tires. Are they fresh at the beginning of the stint or are they worn and losing grip? The following driver may decide to drop six to ten inches to the inside; that's just enough to get a little air on your left front wing. This will also affect the left side of the rear wing, plus you get some air to your tunnel. We create a lot of downforce with a Venturi-type effect in the tunnel. If the guy ahead of you goes down low, you do the opposite with your little move to the right. You're always playing with that—how close can I get? How much of a "washout" can I take? How much do you need to sidestep the person in front of you one way or the other?

Washout? (I'm just using arbitrary numbers.) Say you have a thousand pounds of downforce on your race car—that is being generated over the wings and down into the tunnels. If you follow a car that is, say, a foot in front of you, you may be creating only a hundred pounds of downforce now. The car becomes less "planted" in the racetrack (you have less grip—nine hundred pounds less grip), and you probably can't get through the corner with a hundred pounds of downforce.

Generally speaking if the car out front is running full throttle, the car that's following will have to "lift" going into a corner. There is another positive tradeoff and often the reason we do drafting: yes, you lose downforce, but you also lose "drag." So if by drafting you now have only a hundred pounds of downforce whereas you *did* have a thousand pounds—and let's say you had five hundred pounds of drag and you now have only two hundred pounds of drag—you are effectively going faster forward, much faster because you have that much less air pressure holding you back. When you're in a draft your air density and air pressure is less. Granted, you have less "hold" on the racetrack, but you also have less holding you back. Your terminal velocity numbers go way up. Your car can now go faster—faster than you could if you didn't have someone in front of you.

There's always a "give and take" in getting a "tow" from the car in front of you. The reason you are able to draft and make passes on cars is because you are your changing your terminal velocity by reducing air pressure. At the same time you are "unsticking" the race car. In the corners you slide more; you're going to slide much more up the racetrack laterally.

When you're in "traffic" you are constantly aware of "dirty air." This air has swirls and pockets—all kinds of things that are being created from the "wake" of the car in front of you. You're thinking, *How much fresh air do I need on this race car to get through this corner flat-out or efficiently?*

A lot of times if the car is not trying to pass the car in front of him he'll regulate his speed by "lifting" on the accelerator or "dragging" on the brakes with his left foot. Doing this on the straightaways in allows him to run flat-out through the corners. This helps with another handling problem: in the corners you're upsetting the balance of the car; you're getting "pitch" and "dive" on the race car. You know that if you jump out of the throttle on the middle of the corner, you're

going to "spike" a bunch of weight onto the front of the car. The car dives, and it will make you loose. That's a mistake that someone who's just being introduced to aerodynamic racing, professional racing, will make sometimes. A panic lift! "Oh, I'm losing control!" and . . . it dives, and you spin. So you're constantly "balancing." A *great* race driver is basically a very sophisticated balancing act. He has a sensitive feel for the race car and is constantly sensing, *How far can I go? How far can I go with pitch and dive? How far can I go with having the least amount of downforce on this race car to get through the corner efficiently? How much do I need to "drop" the car into clean air to get the car through the corner?*

Passing a car in traditional drafting is to "slingshot." You want to have lower drag numbers to get a running start at somebody, so that when you do pull out and subject your car to the pressure of the fresh air, you have enough momentum to get past the car in front of you. There's nothing he can do about it because he's flat-out, other than to get right back and draft you and slingshot past you.

For me it's a matter of timing; I want to time the slingshot just right. There's a race in the balance. The guy in front has it on his mind: is there any way to impede, without blocking, without being "dirty"? The Indy Racing League does a great job in regulating fair play. Their ethics of racing are very clean and as safe as they can be in a very dangerous sport. So obviously in a "no-rule" race, the car in front of course has mirrors and can see that the car behind is making a run by getting a slingshot. You can see that the car behind you is going to pull out to the inside; if there were no rules you'd pull to the inside to impede that.

The IRL rules: you can't impede the progress of the car behind you. If you're running a "low" line you have to stay low on *entry*, in the *middle*, and on *exit*. You can't be weaving back and forth. That's not allowed in any form of motorsports—weaving and blocking. It's more than the gentleman's agreement; it's enforceable by the rules.

The bottom line is that everyone competing at this level of racing is a professional. There is a professional, mutual respect. That's paramount, because what you do to one person they're going to "do unto you."

Now, if you know that car behind you—by watching him, by knowing how that driver "sets up" or how his car is working—if you've done your homework through the race and you know, perhaps, this car you're racing with is a little bit on the "loose" side at one end of the racetrack, you know that they have to enter the corner at a certain "line" because of the balance of the race car. So, ultimately, if you want to make it the hardest for that guy to pass you—for the win of the race—then you'll pick a line that is most advantageous for you and not advantageous for the guy you're competing against. It's like in boxing where you're studying your opponent for the knockout punch; the key is *timing*. In racing it's the same way, if you're studying the people you're racing with. If you're going to have to make a pass to win the race, if you've done your homework, you've studied their trends. A lot of times you come into the pits, you may have had a loose race car and you get the balance just right for the end of race. Now, maybe you can do certain things that you couldn't do before, running different "lines" on the racecourse.

Some guys' cars, by the end of the race, can only run on the bottom of the racetrack. It's just the way they've set them up, the way their setup has filtered through to the end of the race. Some guys' cars, by the end of the race, can only run on the top of the racetrack. Then there's somebody who with that magic driver/engineer combination set the car up *perfectly*. They can run on the bottom, the top—anywhere they want to go—so they can make passes much easier than the other two types of cars. There are cars that were set up with less downforce, that are running less "wing" so they can make passes more easily, but they're not sticking as well through the corners. Then there's the car that's running more downforce; it's harder for them to make a pass. They have to really time it right because they have so much more drag in the car, yet when they get to the corner they're just "glued." They can run high, low—anywhere they want to run.

You know that by the end of the race that your competitor is trying to fix that weak spot. That last pit stop he may have made an adjustment to fix it. But was it enough? A lot of races have been won because a car was handling a certain way and they made a huge change during the final pit stop. Everybody that thought the car could only run low on the track finds out that they can run high and can pass them easily.

A driver always has a natural propensity to find trends. You're always trying to find engineering trends to make your car better and better and better. During the race you're watching for trends in your own car as well as that of the competition. Drivers at this level all have the knack for watching the trends of the entire field.

There are so many engineering factors that go into the setup of a race car. We want to know what that car can do on race day. That's why we test and practice, test and practice on the racetrack and simulations on the computer. There are experts, engineers, and aerodynamicists sitting down twenty-four [hours a day] and seven [days a week], 365 days a year doing these computer simulations. The race driver has to be very good at taking what he has, communicating to make it better, and then *driving* it. You drive whatever you've got on race day. Communicating before the race, during the race, after the race, after test sessions with the engineers (and hopefully those engineers have the skills to take that information and turning it into a better car within the rules)—it's a very difficult process. Generally who wins the race is the best driver/engineer team who not only did the best job on race day but did the best preparing and made the best changes as the race went on.

If you're winning 30 percent of the time you're phenomenally successful. Even 20 percent of the time is amazing, but you have to be able to deal with the fact that you are losing 70 or 80 percent of the time. With the team there is hopefully a season-long "migration to perfection." That's why the end of the season [is] so competitive—most of the teams have "mastered" the race car.

Then everything gets recalibrated, and you start the next year.

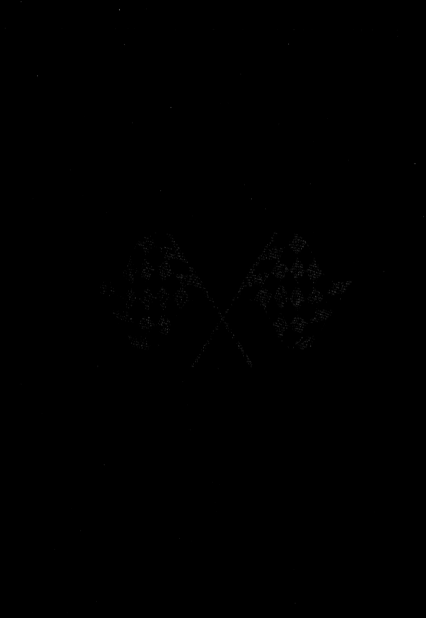

# 17
# TOM SNEVA

*Won Indy 1983*
*Date of interview: February 2004*

Tom's portrait is the only one shown on the trophy wearing eyeglasses (by his own request).

## Q: WHAT DO YOU THINK OF THE WAY THAT THE SPORT OF RACING IS EVOLVING?

I think racing today is good from the "sports/entertainment" side of the field. The trouble with trying to be in the "high tech," "leading edge," and all that is that it's cost prohibitive. It gets the costs so far out of sight that it makes no sense. Also when you do that it's hard to get the competition to be close, because you've got guys with unlimited budgets, and you've got guys with little budgets. If you let technology go, with no restrictions, then the guy with the most money is going to win, and he's going to win by a lot. Therefore it won't be very entertaining to watch, and therefore the sport can't grow. That's my opinion, anyway. No question that the IRL has hit on the right formula. But there's a catch-22: what drivers are going to like and what's entertaining won't be the same thing.

I know that when I was driving I wished that everything would be pretty equal, and then my driving ability would make the difference in how we did. I think that would have been good. If they didn't have the downforce that they do today . . . the guys now on most of these tracks just run "flat-footed" around the racetrack. Nobody has to "lift," and that takes the driver out of it somewhat, but it makes it very entertaining. Everybody is running pretty much the same equipment, and it causes the races to be very, very close when you don't have to lift in the corners. The problem is when you get to racetracks where you *do* have to lift in and out of the throttle and use the brakes and stuff, then they tend to get a lot more spread out. From my standpoint I wanted the equipment to be pretty much equal and not have so many aerodynamic rules that let you drive flat-out, so the driver would still be a factor. For me, I thought that was the best of all worlds when I was a driver.

But for the sport—and to be entertaining— you've got to get on these tracks where they can, pretty much, run flat-out. It makes it really exciting to watch because they don't get spread out too far.

TV is the only way you can grow it. That's how other major sports have grown—making it popular on TV, getting the TV money, and the things that go with that. If you're on TV, then there's companies [sponsors] who want to be involved. You have a lot more potential for growth.

***

I don't think speedways ever become obsolete. Tony George and the Indianapolis Motor Speedway have been in the forefront with these "softer walls" things. He's trying to "cushion the blow" a little bit for the higher speeds.

I can remember, I was the first person to run 200 miles an hour at Indy in 1977 qualifying. Pete DePaolo, the 1925 esteemed Indy winner, was alive at that time; he was

Sneva signing autographs at a Texas event.

the first to go a hundred miles per hour at Indy. He even related the story to me—he remembered talking to reporters after he ran "100" and they asked him if he thought they'd ever run two hundred miles an hour here and he said: "You'd have to be mentally insane to even think about two hundred miles an hour at Indianapolis." Yet he was still alive when I did it.

When I look back on it, those guys were crazy. They sat out in the middle of nowhere, with little teeny tires. You know, back then they had *skinny tires* and *fat drivers* and now we've got *fat tires* and *skinny drivers* [*laughs*].

I know about Indy. Any time you had a racetrack that was all bricks, it had to be rough!

But it is amazing. Now the cars are so much better. With the rules the way they are now at Indy, qualifying, it's just flat-out put the pedal to the floor and steer it around for four laps. That's why you now see rookies doing a lot better than they did years ago. There's not so much "in and out" of the throttle where you have to . . . where the driver's ability is a bigger factor. *But* today it's a lot more entertaining.

Part of your job as a driver—it's kind of like pitchers and hitters in baseball—you have to have a mental "book" on everybody you ever raced against. Like hitters have books on pitchers they've faced and vice versa. You've got to have that mental book and know where their strengths and weaknesses are. At Indy you've got thirty-three of the best, but you've got a wide variety of abilities. It's up to you to know. Some guys you can run wheel-to-wheel with in corners and some guys you only pass on straightaways. But that's part of your job, and if you don't do it well, you might not be around that long.

You could go right down the line with everybody. A guy like Rick Mears you could run side by side with him on any part of the racetrack and not be concerned about it. A guy like Mario you had to treat a little differently. He was very competitive. He was probably the toughest guy day-in-and-day-out to try to pass. He'd use up all of the racetrack and do all that he could to intimidate you on the racetrack. So every guy you faced you had to keep a book on.

Psychology is a big part of being competitive and winning. Probably the king of psychology was A. J. Foyt. Foyt was the best at getting his competitors' minds to think about things that weren't as important as the things they *should* be thinking about. I remember one year, he was still real competitive probably in the late seventies. He had a leather cover that he put over the nose of the car. It was always the first thing he did when he went into the pits—he threw this cover on—and it was always the last thing to come off when he exited the pits. He was still real competitive so everybody was concerned about his "speed secrets" and what he was doing. I can remember guys like Bobby Unser—they'd hire photographers to try to get pictures of the nose of the car when it was on the track and try and figure out how he was going so fast. This went on for most of the season. One of the last races that year was at Phoenix. Before qualifying, in the pit line, he brought everybody over. It was a big "grandstand" thing. He brought all the Unsers, Andretti, myself—he brought everybody over—he took the cover off the nose of the car and there was absolutely nothing there. All of the "tricks" were on the *back* of the car, but he had us so focused on the front that nobody *looked* at the back. So he got away with that for a year!

The key to these things is that you've got to learn from those situations.

I remember, I wore glasses when I drove, most of the time, and there in the mid to late 1970s we were always qualifying "pretty quick." Well, I have different glasses for different situations. I had a real bright red pair of sunglasses I'd use for qualifying. Then I had a real *thick* pair that I'd wear to the drivers meeting. I'd walk into the drivers meeting and maybe trip over a chair or two with these thick glasses on, seeing if I could maybe get their attention and they'd give me a little more room on the racetrack that afternoon.

[*Laughs*] These are all the things you do to get these guys' minds on something else instead of the job at hand. They're little things and they seem sort of ridiculous at the time, but every little bit helps. It was so competitive that you did everything you could to try to gain a little advantage.

## THE AERODYNAMIC "TUNNEL" UNDER THE RACE CAR:

Basically, you're changing high pressure into low pressure. Nobody understood that—what the air was doing underneath the car, between the racetrack and the bottom of the car. There was a lot of high pressure forming that was actually trying to "lift" the car. Then somebody figured out that if you shaped the bottom of the car a certain way, you could turn the high pressure into a low pressure and get the reverse of a wing-type situation. Instead of causing lift, turn lift over and you've caused downforce. It's pretty amazing—everyone was concerned about the shape of top side of the car for so many years, making it look aerodynamic, but we gained 200 or 300 percent more efficiency when we started paying attention to the bottom side of the car as opposed to the top.

Jim Hall was the first to bring the innovation that we call "ground effects" to Indianapolis. That was [in] 1980 when Johnny Rutherford was running what they called the Chaparral. It was the first "ground effects" car to come to Indianapolis. It was just head and shoulders better than everybody else at the time. He was the first one there. I think Colin Chapman brought ground effects to Formula One in the late 1970s. In fact, one of the years that Mario Andretti won the Formula One championship, that was in one of the first ground effects-type cars. He was the first one to apply that technology to the bottom side of the racer. I'm not sure who the engineer was to develop these ideas initially, but I know that Jim Hall was the first to use it in an Indianapolis car that Johnny Rutherford ran.

Rutherford won that year. I finished second. I started dead last in a backup car, a car that was two or three years old. People ask me what my best memory of Indianapolis was. Some people think it would be the first two-hundred-mile-an-hour lap in 1977 or the win in 1983. Well, actually it was running second in 1980, which was actually the third time that I'd run second! To the average

spectator, he might think, *God, the guy can't* win *the race. All he can do is run second.* But we started dead last with a car that was three years old, with a team that was under-budgeted—just a bunch of renegades. We just maximized an effort, with the crew, in the pit stops—me and the racetrack. That was the first year that I actually brought spotters to racing. Right now if you look at the NASCAR races, half the time they're talking about the spotters—the people up in the grandstand trying to help the driver get around the racetrack. I brought that to Indy in the late 1970s.

You know, before I started driving in the "big leagues" I taught school and was a football coach. Back then, we used to have assistant coaches up in the press box. They had a better view of the field and we could talk to them about "formations" and stuff like that—what's going on. On the sidelines you've got a limited view. So I took that theory to Indianapolis. Back then we'd talk to the crew chief who is in the pits. And you know at Indy when you're in the pits you're down low—all you see is the cars going by at two hundred miles an hour. It's sort of a blur. So I decided that we needed to put people up high who could see other parts of the raceway except the front straightaway. Not much trouble happened after the start of the race on the front straightaway. It became a big plus for us and gave us a big advantage. In 1980 in that race we took advantage of all those techniques. So to me, that one day was the most rewarding at Indianapolis.

They began to catch on! Now everybody has a spotter—more than one. Back then there was only one or two of us. The teams all had their own radio frequencies, the safety workers and emergency people had theirs, as did the spotters; and the radio communication wasn't that great. I can remember, I think it was 1981, back then we had some unwritten rules. We'd say, "Crew chief, you can only talk to me when I'm in a straight line." Don't talk to me when I'm about to enter a corner or something like that 'cause we're focusing at two hundred miles an hour, trying to hit the apex, and if we hear somebody come screaming over the radio at an unexpected time it can upset your focus. So we had some unwritten rules. A lot of times the mechanic would get excited and forget 'em, but then, also, we'd have a lot of "cross frequencies." You'd start getting intermittent communication or start picking up somebody else's frequency.

I remember in 1981, we were just a little bit into the race, and I'm getting a lot of cross-frequency chatter over the radio. It took me a few laps to realize that it was one of the observers. They have twelve observation posts around the racetrack. The guy at "post nine" had spotted a good lookin' gal in "Section 23 - Row WW," and he was trying to tell his buddy over at observation post eight where she was. I was in a pretty good battle with Foyt at the time, and it took me about twelve laps to find her myself. [*Laughs*] These are some of the stories that don't always get out! These are some of the problems that drivers have to overcome, that fans don't realize, at times.

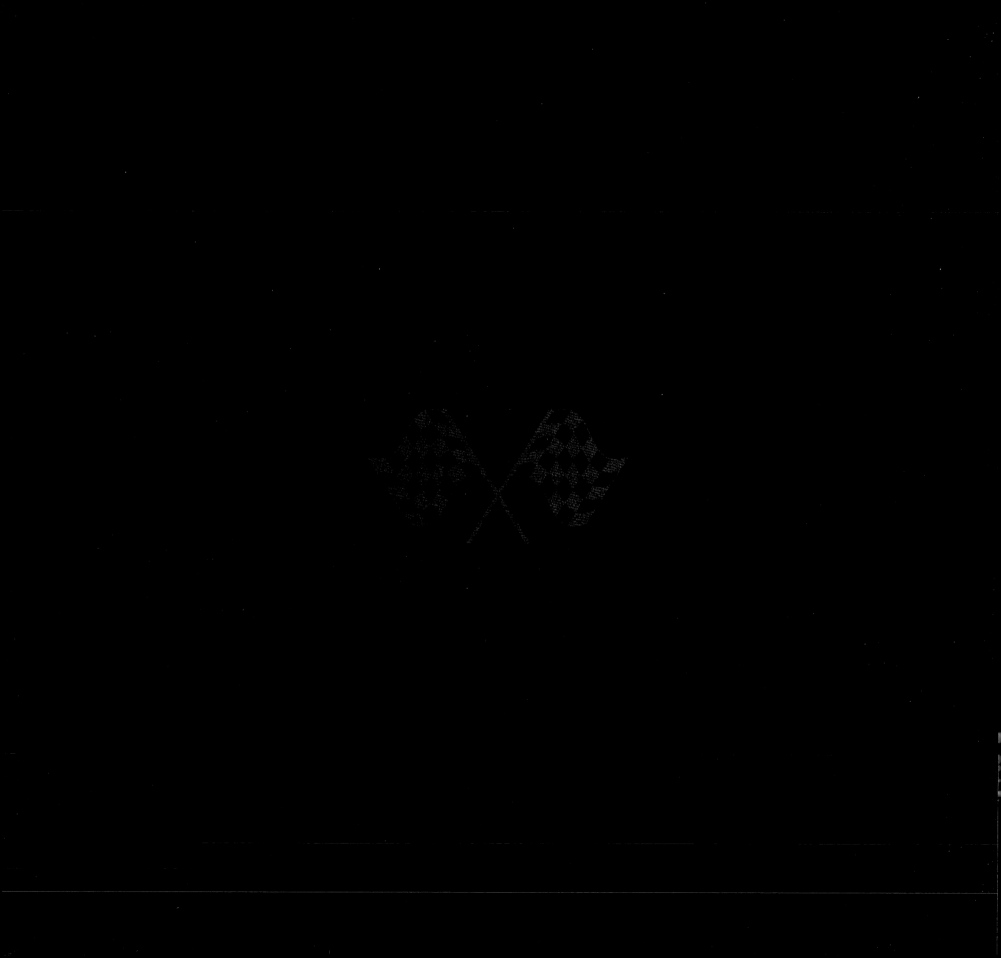

# 18
# PARNELLI JONES

*Won Indy 1963*
*Date of interview: February 2004*

It wasn't that easy to drive, but it turned out to be very good for the race. Not so good for qualifying. Anyway, I got a lot of money for driving it so I thought that was a good thing to do at the time. It turned out to be a good car, but I abused it, probably a little bit too much during the race. A bearing went out in the rear end. It cost us. Three and a half laps from the end it quit.

In 1968 we ran again—with a restricted version—I didn't drive. That's when I quit . . . after the '67 race. I quit running "open cockpit" cars and stuck to TransAm, stock cars and "Baja." Anyway, in '68 again, Andy Granatelli had Chapman build four of them, and I guess they ran four of them at the Speedway.

Driving it had a *three-second* "throttle delay time"; you had to kind of outguess it.

You know you get a feel for it pretty soon; you just had to keep concentrating on it. Also when you back off at the end of the straightaway, you didn't have any compression to slow it down; it was as if the throttle was stuck, so you literally had to use the brakes to make it slow down. We had adequate brakes on there. Of course you don't slow down too much, [but] you better slow down *some*, or it's: "Hello wall!" [*laughs*]

## Q: DID THE TURBINE MAKE A LOT OF NOISE AND WAS THE EXHAUST BOTHERSOME TO OTHER DRIVERS?

Well, they tried to make big stinks out of that in practice. It was a lot of fun because I was needling the other drivers. I'd say, "How fast are you going to run in the race? I wanna set the screw."

Of course I didn't know what I was talking about! I knew there was a little screw on there somewhere that made a difference, so I was needling Foyt and Andretti and some of those guys. You know, I was always in the first two rows when I was driving at Indianapolis. I was sixth fastest—second row on the outside with that car. But I knew, come race day, when they took their seventy-five gallons of fuel that we were carrying at the time, when they put on seventy-five gallons of fuel and took out that 10 or 15 percent "nitro" that they were using for practice and qualifying, probably, I'd drive away from them. Basically, that's what happened.

The car handled well. In the first turn I passed all of them along the outside— except for Mario, and I caught him . . . coming off of two and I passed him coming out of two. I took the lead in half a lap. I said, "Good-bye!"

It was a helicopter engine, a Pratt & Whitney ST-6B gas turbine. It had a one-speed transmission. That's all I know about it. It only had about 550 horsepower, but it had an awful lot of torque.

## Q: DID THE ENGINE BURN FUEL FASTER THAN THE OTHERS?

We made about the same number of pit stops. We didn't have much of a problem there; I mean, we carried seventy-five gallons of fuel!

## ON THE DECISION TO QUIT OPEN COCKPIT IN '67:

I was fortunate—my business was taking off the ground after winning in 1963. The turbine quit right near the end of the race in '67. I was thinking that winning again was not going to be as exciting as it was the first time. And then when the car quit, it helped me make the decision. Gee, if it was not going to be that exciting, and my financial future was really in my business, I thought, *Why are you doing this? So, maybe it's telling you something.*

I quit driving open cockpit cars and went to TransAm, stock cars, the Baja races—things like that (something with a "roll-cage" over me).

I had started a chain of tire stores. I don't remember what I had at that particular time, but I built it up to forty-five stores. We had three wholesale warehouses. I was growing with that business at that time. I was also a "race tire" distributor for Firestone. We still make a tire with my name on it, a recreational tire, an off-road tire. It has a "PJ" in the tread. It's going real good.

## TELL ME ABOUT THE THINGS THAT ASSAULT AN OPEN-WHEEL RACER:

My rookie year something hit me above the eye. It started bleeding, it bled for a little while, and then quit. When you're young . . . and you don't know any better . . . racing? Man you think you just died and went to heaven. You think this is the best thing that ever happened to you. Certainly it was a lot of fun . . . even though you know that it's very dangerous.

That certainly stands out. Once you've won the Indianapolis 500 it kinda puts you in a stature or some position: "Hey, you're an Indianapolis 500 winner." And it does open the doors for you in many ways. For me, it's the most exciting thing that's ever happened to me. Also, I was the first to run 150 miles per hour, and that was truly thrilling. They'd been trying to do that for two or three years. [American driver] Jim Hurtubise did a run of 149 two years before. Everybody felt that that was a real "magic number." So to be fortunate to do that was thrilling at the time.

I had learned something from my rookie year. It was something that gave me about a mile-an-hour edge. You know, most guys set the car up so it understeers. The reason for that is that the tires that we were running at the time were very narrow. Real hard rubber. They didn't hold very well. What happened was that if the car got loose on you, they used to say, "Go ahead and spin it out!" So, everyone was afraid, all the time, to run the car real loose. If they did, they came in and corrected it. There was a guy—Jack Turner—who flipped down the front straightaway at Indy two years in a row. I watched Hurtubise when he qualified that 149 miles per hour. He and I were brave and not too smart. I watched the film of him and I noticed the back end of the car was slipping a little bit, like he was a little bit on the loose side.

So, the first year, in the race, I almost spun the car. I was barely able to correct it and keep it out of the wall. But I knew then that the next time I got that car a little bit "sideways" I was going to be ahead of it. I'd figured out what it was going to do. Later—not that year—during testing I started working on that: *slipping* the car. I found out I could get away with it. Boy, that really picked up the speed. What it did was make the car come off the corner a little bit without being "bound up." It came off a little freer. The competitors didn't really catch on; a lot of 'em didn't want to try it. But I didn't tell 'em what I was doing either!

Guys sometimes "drive over their head." It takes no talent at all to run down there and hit the wall, so you've got to be able to drive that car on the verge of losing control of it. That's where you've probably heard the words "car control." Some drivers have great car control

(some more than others). There's even more than that. It takes a lot of will to win. You have to have the determination, the desire to win. I occasionally go and participate with some of the driving instructors at the driving schools. You can teach somebody to drive, pretty well. But you can't teach 'em that desire, that will to win. You've got to have that on board. I can spot it in a prospect in a lot of ways. I saw it early on with Jeff Gordon, Tony Stewart, and Robby Gordon; they have car control. Mario Andretti showed it early on. A lot of race drivers . . . most of the top drivers have that talent.

It was a lot different then. Today, there are so many forms of racing. It costs so much more. You have to have sponsorship to make it work. You know what I mean? Before, you could have a car owner who could afford to race without a sponsor. If you got a sponsor, that was fine, but they hired you more for your talent. Today, a lot of our racing is based on who can bring the most money.

Take a driver, for instance. Say he's an A-OK driver but he's not outstanding and he's looking for a "ride" with a car owner. Suddenly a driver comes in with seven or eight or ten million sponsor dollars behind him . . . what are you going to do? You have to look at it as a business.

Racing has been hurt that way, in a lot of ways. Aside from the financial problem there is also the fact that it's aerodynamic. You've got to have aerodynamic engineers. Electronics have hurt it a great deal; it's hard to control the electronics in cars. You've heard the words "traction control." A while back, with CART, they were running traction control, but nobody could pass! Part of driving is being able to "feed the throttle" an accurate amount to keep from getting "wheel spin." If you can just reach over and mash your foot to the floor and you don't get any wheel spin, nobody makes any mistakes! That's one of the reasons that they tried to put that away. And when you have electronics, that's pretty hard to control.

Launch control! That's what I'm talking about when I say electronics is screwing up racing. NASCAR still has carburetors, but when you have "electronic fuel injection," "electronic ignition systems," it's hard to control.

A rev limiter is a different deal. That's just a matter of keeping 'em . . . making the engines more reliable and not turning 'em too tight and maximizing the power. I mean so that everybody is reasonably equal.

# 19
# ARIE LUYENDYK

*Won Indy 1990, 1997*
*Date of interview: February 2004*

THE THING about driving in Indy is that when you first go there you're intimidated by the size of the track. Even if you've never driven there, you know what's happened there in the past. I started in '80 and was aware that there have been a lot of big crashes there. Mario Andretti gave me this advice, he said, "Don't rush into things or the place will bite you." I always remember that, and I've always treated the place with a lot of respect. Every year I went back, I always remember to get comfortable with the car before I "went for it." I was comfortable with it after the first year. It doesn't take a race car driver that long to get used to a racetrack. If it takes you longer than a year then you're in trouble; you're probably not doing your job.

As far as Indy, it's not that hard, but to race five hundred miles, to be able to see the end of the race . . . that was hard. What I did in my first race, I kind of "drove my race" and went from pit stop to pit stop. As I progressed through the race I got more and more comfortable with the race itself and myself in the car. Toward the end of the race I began to feel really, really comfortable. I was very happy with the results that first year. I finished seventh. I was rookie of the

Arie and Arie Jr.

goes hand-in-hand that you have to bring support, sponsorships, and money with you. When I started racing in the United States in '81 I set my goals for the IndyCars. I raced at PIR—Phoenix International Raceway. That was my first race here in the States. I found a company in Wisconsin that sponsored me called Provimi Veal. They had an IndyCar team that Tony Bettenhausen drove for. I started to drive in the Super V series. After I won the championship in the Super V series they finally let me drive their IndyCar in one race. That's when I really got involved with the Provimi VealCompany. They really helped me. The owner of the company was Dutch. Without his sponsorship and his help I never would have made it into the IndyCar series.

IndyCar racing became really popular in Europe when Nigel Mansell came over. It made the sport more international. Then you had Zanardi and a bunch of guys from Brazil. At one point you had more foreigners than you had Americans.

I think Americans, in general, have always welcomed foreigners into their sport, into their country, and into their life. There may have been a problem for American *drivers* when all those guys came over. The Americans had been mostly racing ovals. When road courses were introduced they were kind of left out. The Europeans and the Brazilians and all those guys would dominate those races, and they wouldn't have much of a chance, except for a select few like Andretti and the Unsers, Jimmy Vasser . . . those guys.

It would've been good if there had been just one open-wheel series.

## MOST VIVID INDY MOMENT:

I guess the best moment was arriving there and driving the race for the first time. But really the *best* moment was *winning* the race and fulfilling a dream. You know I did think about winning that race, in my mind, many times before I actually won it. I always had the feeling I could win that race. With the right people surrounding me it could happen. And *when* it happened it was the accomplishment for me that I was striving for. So that was awesome.

## ON POST-RACE VICTORY ELATION:

You know what happens is that it's such a monumental moment. You want to savor the moment for a long time. It puts you in an

year. After that first race, the way I came out of the car I knew I could be successful there. I always strive for that and I did become successful there. And success there is not all about the driver; it's really about the team you're with and the equipment that you have. You have to get the total package right.

My first five-hundred-mile race was Indy. I had only done three IndyCar races before I drove at Indianapolis. I think I've raced there eighteen times. I'll have to go back and check the record books.

I grew up in Holland, and my dad used to race Formula Vs and Formula Fords. So I kind of evolved into racing that way, working in shops when I was younger. When I was eighteen I went and got my racing license and began racing. I had a long career in Europe with several types of racing—open-wheel racing. I won a bunch of championships. I never really made it to Formula One because it extremely . . . good kind of mood. You're just pretty relaxed after that. You want to take your time . . . and all that. I was doing interviews forever. That day never ended.

The second time was pretty much the same—[just] as enjoyable. I mean for many years I was pretty close to winning the race. In '97, when my second victory finally happened, it was pretty much of a—you know, relief.

I know some drivers, like Mario Andretti, he won it once and never won it again, and he's been pretty much chasing it for many years. He had a lot of bad luck. I had bad luck too, but finally in '97 it came together again.

## 20
# BOBBY UNSER

*Won Indy 1968, 1975, 1981*
*Date of interview: February 2004*

Bobby Unser proudly shows off one of his championship checkered flag rings.

**PEOPLE HATE** technology; they really just want to see a contest between human beings. It's a fan-based sport. It doesn't make its money by just corporate [backing]; it's [from] fans [too]. Therefore, if you don't write about the people that make it you're not going to have anybody read your stuff.

## ON THE CRAFT OF RACING:

Drivers are all a little bit different. Some of them have to psych themselves up to go fast. But I would like to think that the drivers that I knew didn't have to do that. They really liked to go fast. The guys that I thought were the best drivers . . . the best drivers that I ever ran against . . . were the ones that truly liked motor racing. They really like to drive race cars. You have a lot that don't really feel that way—they like the money, they like the girls, the excitement, the glory . . . they like all that kind of stuff, all those things that they don't get in ordinary life. But a real race driver, a true race driver, really races because he likes it. Money isn't the most important thing, but success is. Have a contest with the other drivers. You can also be friends with them. In my era of racing I had a lot of good friends that were the finest race drivers that I ever knew. Yet on the racetrack it was bitter, bitter, bitter to go against each other as hard as you could. It became an out-and-out outright contest. Man against man. You knew the guy's habits—all of his habits. You really got into your racing.

It's a little bit different today. I'm not saying that the drivers that make it today aren't good. The ones that make it to the top today are good; in fact, they are extremely good. There's just no question about it. But if you take the drivers back in our era—my brother, A. J. Foyt, Johnny Rutherford, Mario Andretti (there's so many of them), Don Branson, Roger McCluskey, Lloyd Ruby—those guys were so good it's unbelievable. Yet they could make their living not from corporations, not from sponsors, but from their earnings from driving race cars, winning, or placing in a race. And that's the total income that they had. Yeah, they did a little PR work and anything they could do, but most of them did very little. So they truly, truly, truly loved to go to a race. They liked it when the green flag fell. Most of these drivers I'm talking about couldn't race often enough. Sometimes these guys (myself included) would race four times a week. That wasn't enough. In fact, I used to lay in bed and dream or wish there was a way that I could run seven nights a week. I truly loved the sport. I liked driving race cars. It was a challenge. It didn't make any difference . . . I never tried to think about making money. I knew I had to make money, but I never let that be the decision. Now some drivers have been more successful than others in the money department and yet good race drivers. Like Andretti—Mario was money-hungry but a helluva race driver. But that's okay. I'm not knocking him. My hat's off to him. I'm glad he made a lot of money; I did fairly well myself. Al did very fairly well, as did Foyt and Parnelli. Think about the guys that did it. They did it because they turn the steering wheel, not because of some corporation that used them for PR work.

Speed is only relative to what you're driving and what your talents are. Someone is always faster than the next guy. Not everybody runs at exactly the same speed. It has to do with the car that you have to drive, the owner, the mechanics, the combination, the team. You look at some years—you look at some of Al's years, for example; in the early 1970s he was *untouchable*. He broke most of Foyt's records because he had a really good combination and . . . they kept the team together. Al

and George Bignotti made a good team; they had good cars and they had a good crew. Everybody has those years if they become famous. I had my years, and Parnelli, Mario, and Rutherford . . . they all had their years when the combinations were right. I had a lot of good combinations, I had good years with Gurney, and I had super good years with Leader Card Racers. That was Bob Wilkie. When I drove for Wilkie, he was the Roger Penske of those days. When I drove for Gurney, we're talking about Dan Gurney, we just made a "super team" out of Gurney's "All American Racers." He was in the business of making race cars, and I was the factory driver.

Dan is actually the guy that came up with the idea of the "wicker bill." I'm the guy that developed it. Dan and I were testing his car at Phoenix. Dan said, "I've got an idea for something to put on the wing." He said, "Would you like to try it?" and I said "Sure." For a workshop he had an open trailer. He went over there and started hammering on a piece of aluminum and pretty soon he comes over and said, "I've got it done." He said, "Let me know when you want to try it" (because I basically ran the test). I said, "Let's try it right now." Dan's a smart guy, always very, very, very good at ideas.

I knew immediately. I didn't have to do a full lap. I already knew that we had a discovery that was beyond belief, probably the biggest discovery that motor racing has ever had in the *history* of motor racing. So I pulled in; I ran just a few laps because I knew we had spies around the track (we always do). The tire people—Goodyear and Firestone—were the big-money people in those days. I pulled in, I didn't act excited or anything, so even the firemen wouldn't see anything. I got Dan and Wayne Leary, my chief mechanic, and I calmly told them, "Guys, we've just made the biggest discovery ever." Of course, Dan, he doesn't understand it. He (honest to goodness) did not understand it, because why didn't I *go* fast?

Well, I was a very smart race driver. I hid a lot of things for many years.

I told him nonchalantly, "Make me a couple for the front." So he took what he'd done to the rear, and now Wayne and Dan and I (everybody pitched in), that's what we're going to work on. At any rate it was a gigantic thing. I'd say it's the biggest single change that's ever happened in motor racing.

Here are two of Bobby's three images on the Borg-Warner trophy—one with a "helmet" and the other without.

We went to Indianapolis with it and I kept a secret. Ironically, I didn't know how we were going to hide it . . . 'cause it sticks out like a sore thumb. Dan's concept was let's just not hide it, because you can't hide it. Don't cover the wings like all these teams do today; keep everything out in the open (he's selling race cars anyway). Sell 'em up the kazoomie, you know? Don't hide them; just leave it. Maybe nobody'll notice 'em. What it looks like was a stabilizer to keep the wing from bending at high speed. With the wind pressing so hard against the wing, it looks like we put a little stabilizer on it to keep it from bending. We used to have the first garage going down the center of Gasoline Alley. I'd stand there and peek behind the curtain because all the cars went by us to go get fuel, on the way to the fuel dump. So I look at all of them going by, the McLarens, all the Eagles we'd sold; nobody copied us—not Rutherford, not Mario, not Foyt, not anybody! So I just can't believe this. I mean I've really, really, really got a big deal. Then I finally saw another Eagle come by. This is a second-class team with no money. They'd bought a new car from Gurney, but they have no engines, no spares, not a good driver, all of that stuff. He rolls by the garage and he's *got* it. He copied it, but he's a real slow car. The car wouldn't go fast if you painted it with a pink brush and put polka dots on it. That thing's not going to help him. He's got no engine. He's lucky if he makes the race.

In other words, they were just a team that didn't have financial capabilities. He was the only guy that copied it, probably out of sixty cars.

That was USAC, not the IRL. We didn't have all the rules that we have now. I didn't have to explain; I didn't have to apologize. There wasn't any violation of the rules. *Nobody* figured it out. Now the whole world uses it. If I wanted to change the size of this "wicker bill" I'd never do it "out there." I'd do that back at the garage. But in the beginning nobody copied us. If we had covered it up I'm sure somebody would've done it.

## OTHER INNOVATIONS:

We once had a turbo charger that was way better than anyone else's. This is in the same era. Ironically I used to walk past Rutherford's garage; I could see that he had one just like it sitting there on the shelf underneath the bench. They never put it on the car.

Today they've got so many rules. The cars are all basically the same. It's a "spec" deal. They've all got the same equipment. The only difference is in the setup. There's only three kinds of cars that are allowed; innovation has kind of stopped. But they do that to limit the speed, for safety.

They'll always have rules, but if they get going too fast it's the drivers, the mechanics, the car owners, the track owners, all these people get together and they say, "Hey, we're crashing too many cars. Maybe we should slow 'em down." Or all the drivers are starting to complain. It doesn't matter to the fans if you're doing 200 or 230 if you're watching the cars. For general safety sake we need to keep the speeds to whatever is reasonable compared to the technology that's making the speed happen. It's technology that's doing it, not the drivers.

Today it's *way* safer. Technology never stops. We are a human race. You can't stop our brain from working. So in order to slow our brains down . . . the "rules makers" keep making rules to prevent technology from benefiting. It just hampers people like myself, who invent technology (speed), to do things that haven't been tried before. That's what our job is.

They've really slowed technology down, at least in the IRL, not in NASCAR. There *is* no technology in NASCAR. It's the same car for thirty years that they've built. Don't get me wrong. I get along really, really well with some of those NASCAR guys, but they've got to be careful or they're going to become a "professional wrestling" kind of deal.

Back in my era IndyCar racing, open-wheel racing, was way more popular. Just when they started having the "wars" [CART vs. the IRL], NASCAR started marketing really heavy: on our side of the street we had "wars."

You know we're not making heroes out of the drivers anymore. It's the cars. In IndyCars the driver is down so low you can't see him anymore. He's got "car body" all around him. It's killing open-wheel racing.

# 21
# AL UNSER SR.

*Won Indy 1970, 1971, 1978, 1987*
*Date of interview: January 2004*

**Q: YOU'VE MENTORED A WHOLE GENERATION OF DRIVERS (INCLUDING AL JUNIOR). WHAT QUALITIES DO YOU LOOK FOR?**

First off I think it's hard for any driver to say what it takes to be a driver. I think it takes in a roundabout way a lot of desire and a lot of ability. So how do you know when a driver has ability? The driver himself doesn't know it until he starts winning. Then everybody sits back and says, "I knew he was going be a winner." You don't know that! I've seen many, many drivers at Indianapolis; they have the ability, but they never figure out how to win. It's so heartbreaking, but once you've won there it changes things so drastically.

A lot of guys go there—Mario Andretti, Michael Andretti, Jim Hurtubise, Lloyd Ruby. Many great drivers have come so close. Mario of course won it once. Many drivers have the ability and they've come so close. Whether it's them, circumstances, or the machine, I could name so many [who] came so close so many times, and you wonder, "Why?"

They ran out of fuel.

A bearing broke.

An engine broke.

Al on the cell phone—allowing us a look at one of his Indy victory rings.

It all has to come together that day. It makes you wonder . . . and I *know*; I've run there so many times. I've come so close to winning there, *many* times, you sit back at the end of the day and say, "What happened?" When something goes wrong don't look at yourself and get depressed. You say, "That's racing!" [*laughs*]. It's better than crying or being sad.

Lloyd Ruby came so close. His fuel man miscalculated once. I think he caught fire a couple of times. It's just sad. You could name others. Parnelli Jones—he's won the race, but he came *so close*, so many times. Mario, he won it once. Why couldn't he make it happen again?

Then again, other drivers have come here and just won it. Then they say: "I thought this place was hard to win." [*laughs*] When that green flag drops there are thirty-three winners heading out, but at the end of the day there's just one.

On the positive side, I once won it on my birthday (May 29). I think it was my second Indy win. My son, Al, won it on my birthday as well.

My father was a car racer but he never made it to Indy. My brother Jerry got killed here. It's a hard sport, but it's something we love very much. We're a racing family. Racing's been good to us.

You can look at many, many drivers, and sometimes—maybe through experience—they figure out how to win. For years they may have looked in the mirror and said, "Am I good?" and the mirror never answers. A lot of drivers have great skills and potential, but it's all about whether they ever pull it all together.

The psychological obviously plays a big part. Your brain has to be involved; it has to function right and make the right decisions. And how does that happen? I don't think there's any training you can do for that. It's natural—you either have that skill or you don't have it.

Being in good shape? You have to be in good shape. And you can't be a wimp. A driver has to take care of his body. You can't get halfway through a race and say, "I'm tired." Conditioning has its positive effect on the mind as well. Many drivers work out constantly. Your mind has to correspond to the body. Your reflexes, your depth perception has to be right. It's got to all work together. You've got to look at yourself. I could tell in my later years that my depth perception was getting bad.

You have to be aware of keeping yourself at 100 percent. I knew when it was time to step aside.

Now I work for the IRL. I do it for fun [*laughs*]. I still love racing. They use . . . Johnny Rutherford and myself to talk to the new drivers to see where their confidence is and help them with that part of the sport. When I was coming up we didn't have anything like that. We'd go to other drivers to try to get some help; some would help you, some wouldn't. Actually, most would help you but only to a point. Then, when you became a "competitor" they wouldn't help you anymore [*laughs*].

Well, it's only normal! Rutherford and myself, we have everything to gain by teaching them and getting them to stay out of trouble out there.

## ON NEW DRIVERS:

A young guy today, first he has to show talent with sprint cars, midgets, sports cars, dragsters . . . .Then you look at how well he promotes himself; to get sponsorships today that's "large dollars," but basically if the guy knows how to drive a race car, then he attracts attention with sponsors, team owners, mechanics . . . .

Al Jr. doesn't yet have a ride for Indianapolis, but I think it'll fall in place. He's very competitive there. He's won it twice. If some team doesn't pick him up, they're crazy.

It's like the last time I won at Indy in '87. I went there without a ride. It was my decision. I had five, six, seven different offers. I turned 'em down, because once you've won there you know that it takes a 100 percent effort on behalf of the car owner and the team. If you have a "100 percent" team, you have about a 75 percent chance of winning. You take a "75 percent" team, you only have about a 50 percent chance. You have a "50 percent" team? What do you have?

I elected to go back there. I didn't think anything that had been offered to me was "ringing a bell" about winning. So I went back there and ended up with a ride with Penske because of Danny Ongais's crash. He got hurt, and there I was!

It all worked. I qualified. I was in the race.

When Al Jr. was running, if he'd qualified on that first weekend, if I didn't have a ride, I was going to go home.

Well, he didn't qualify.

I had stayed there to help him.

It was "storybook gold." It was a storybook race.

I ended up winning the race.

Everybody said: "I can't understand this. It's impossible!"

## 22
# AL UNSER JR.

*Won Indy 1992, 1994*
*Date of interview: January 2004*

GOSH, THE way I was raised was "wanting to be like my father": I think that's every son's feeling. My dad started me go-kart racing at nine years old. From that time forward he really just taught me the value of life. He taught me the instruments of racing to teach me about life—good sportsmanship, you have to work hard every day to advance yourself, to be a better person. You have to work at it every day of your life, and as soon as you don't, someone will come along and do it better than you. He used racing to teach me all those things. It truly is something you have to work at every single day. You have to think about it every day in order to become a winner and a champion and all those things. It's a wonderful business. Also, you know the saying "what goes around comes around"? If you do someone wrong on the racetrack he may not have an opportunity to get you back, but believe me, he will get you back. So you have to treat your fellow competitors fairly and honestly.

## ON BEING A SORT OF TEST PILOT:

It's a privilege to get to do something that most people rarely get to do. We used to "test" quite a lot and that was when the rules were not

as strict as they are now. Both of my Indy wins were in cars where I was the "test pilot." I had teammates with both of my wins, but I was what you might call the lead test pilot. The first car I won with had a chassis called the Galmer. The only other person that had that car was my teammate Danny Sullivan. We were lucky to win the race that day because Danny had a really bad day. My car ran really, really well. When we first built the car—on its very first run—I was the one to drive it. As I say, it was an honor and privilege to be the first one to drive. That story was more about the car: that is to say the

chassis, because the Chevy engine was a known component. With the Galmer, it was more "the car" than anything else—the gearbox, the rear suspension, the tub, the aerodynamics, the front suspension. Everything was an unknown, and it had never been driven before. I *love* developing cars. That was my tenth year running at Indy. I had just turned thirty when I won my first 500.

I had developed cars before that. At my very first rookie season in 1983 I was in the Eagle—the 1983 Dan Gurney Eagle. It was developed in combination with Rick Galles, who was my car owner,

If it has wheels, Al Jr. will drive it.

and Dan Gurney, who I'd say was the "owner of the engineering." He was a pioneer. He's been "banging on metal" his entire life.

I'm really proud of the years I spent with Roger Penske. The engineering and the dedication that he and his entire team put forth to win all the Indy 500s, the national championships—Roger is truly amazing. My very first year with Roger was the year I won Indy for the second time. Roger started as a driver but moved into being an owner very quickly. He developed and drove the early Corvettes at the Sebring Twelve-Hour Endurance run in the early 1960s. I liked the way that he ran the team and the way the team operated. I told you that the first year we won Indy it was the *car* that we developed. The second year I won Indy it was the *engine* that I developed . . . or I was the test pilot for. I drove it for the first time. It was a push rod engine. It was experimental and quite frankly developed in a record amount of time. I think in a span of about three or four months this engine went from being on a piece of paper to being a race car and running. The manufacturer was Ilmor. It had the Mercedes label on it. That was my first year with Roger. He built his own cars for a very long time. I was with Roger for a time period of six years. My first year I was involved in the development of what we called the 209 (the 209-cubic-inch Ilmor / Mercedes Push-Rod Engine). It won the Indy 500; it was so powerful [*laughs*] that by midsummer they put sanctions on it—they reduced the power in it enough to where it was no longer a viable piece that could come back and compete. They took the guts out of it. That's how strong this engine was. It ended up as a "one year one off" kind of thing.

The 209 engine was so powerful, we'd come off the corners and you could feel the car accelerate all the way down the straightaway. It would just press you back in your seat. You'd stand on the accelerator . . . that car was a beast. It had close to a thousand horsepower. It had a really low rpm, which made it a "thumper." It made it just moan. It was like *w-o-o-o-o-h*. It was a thrill of a lifetime.

Roger Penske took Detroit Diesel (which was a failing business), and he took an existing engine (which was a straight six diesel engine), he put an inter-cooler and a bigger turbo on that engine and totally turned that whole company around. The "Series Sixty" was what he called the new engine. I've got one in my motor home and it runs super. Roger has a commitment to detail and the sheer desire to be the best that he can be.

There's young kids all the way to the old engineers that are involved in designing engines. It's the art of invention. Trying to make tomorrow better than today—that was their field. My field was "How do I keep it wide-open through that corner, where no one else can?" That's my field. With an aerodynamic engineer he's dreaming about, "How do I make that car stick? With the 'air'? So my test pilot can go through the corner." Then you've got a mechanical engineer who is building a suspension so the entire contact patch of the tire, when it goes through that corner, it rolls. The tire rolls on its rim. How do I make that suspension flex and have my aerodynamic engineers still be happy? And make the car stick through the corners. And then you've got a mechanical engineer of the engine. His way of making tomorrow better is, "How do I make that engine run wide open and produce the most horsepower, the most torque for a long period of time, and get good fuel mileage . . . and have reliability?"

You have to be physically fit, and you have to be mentally fit to get into these cars and push the envelope where not only are you on the edge of the envelope, but the only thing hanging on to that envelope is your fingernails. You've got to keep from going over the edge and off the cliff. That means losing control of the car, hitting the wall, and bending. That's not what you want to do: what you want to do is to take it to the limit and find out its weakness. You have to be willing to go to the end of the envelope, and you have to be aware of everything that's going on around you, meaning: What's the car doing going down the straightaway? What's it acting like? 'Cause every car I've ever driven have different personalities. My best cars have been "girls"; my worst cars have been "boys" [*laughs*]. You know its gender right away! If it doesn't go straight down the straightaway, you've got problems. You've got big problems. You are not going to know until you get in it and drive it out of the pit box—drive it up to speed and see what it can do. As a driver it all comes down to confidence in the machine that you are in. What gives you that confidence is when it responds to your thoughts and your movements. It responds to the changes that you make to the car.

Keeping a driver cool when he's already suited up, strapped in, and waiting to take a run is always a priority. Al Jr.'s crew member is shading the direct sun and a fan is directing a bit of a breeze towards his face and helmet.

The whole thing about getting it through the corner wide open is balance. When you take it in there and turn the car and the front end, the tires, they don't grip the ground and make the car turn. You feel the front end slide. That is called an "understeer" or a push. What would fix that? We'd say you want a "softer front end." You'd want more weight on that front tire to make it turn. You come in and tell your engineer that's it's pushing. One of the first things you want to do is soften up the front end, and when you do that and you go out and it makes it turn better, that instills confidence in the driver. You have a predictable car. That builds confidence. If you soften up that front end then go out and it does something that you didn't expect at all, it may even push *worse* or it may just turn in way too quick. You now have way too much front grip; you now have a loose condition where the rear end is sliding and the car wants to spin out with you. So you're forever searching for the perfect balance in a race car as it's going through the corners. It's true on oval races, road courses . . . the only place it isn't true is drag racing, because they don't turn [*laughs*].

You soften it up with either an anti-roll bar or a spring. For example, statically, our cars weigh 1,550 pounds. By rules they cannot weigh anything less than that. When it's running two hundred miles an hour it's producing over three thousand pounds of downforce. So it's producing twice its weight as it sits in the garage. Our front spring is in the 2,500-pound area, sometimes three thousand pounds. I've actually had six-thousand-pound springs on my cars in the past. In other words, you squeeze the spring one inch. It takes six thousand pounds to squeeze it that one inch. If we talk about a five-hundred-pound spring, it takes five hundred pounds of force to squeeze it that one inch. That first inch is what the engineers are concerned about. I've gotten into arguments about what I call a "rising rate spring." Yes, it takes five hundred pounds to squeeze it that first inch, but the second inch may take a thousand or twelve hundred pounds to move it.

The driver has actually three things that he can adjust. He can make changes to the anti-roll bar, can adjust the rear anti-roll bar, and he can adjust what they call the weight jacker. That's a moving platform on one of the corners of the car. It's operated hydraulically. It might be front right or left rear, [or] it could also be front left or right rear. It's up to the team.

The front anti-roll bar and the rear anti-roll bar are controlled by levers that have notches in them, numbers one through five. In the number one slot you would be "full soft" and the number five slot you'd be "full stiff." Nowadays there's a readout to tell you what slot you're in. In the past, you just had to look down there [*laughs*].

The weight jacker is a button. It controls cross weight of the four corners. It's useful on an oval track, but in road racing we just run it straight up because we turn the car both ways. With an oval you're only turning the car in one direction.

Imagine the car as it sits on the tech plate. Each tire is on a scale [to] add up to the total weight of the car. It's diagonal: If the jacker adds thirty pounds of right front weight, *usually* you'll have left rear weight. If you add thirty pounds of left front weight *usually* you'll have a right rear weight. Imagine an "X." Now these are rear-driven cars. In other words, the drive wheels are the rear wheels.

We run what we call "stagger." If you take a tape measure and measure the circumference of the right rear tire, it's bigger than the left rear tire. The tire is actually bigger. Your "fronts" are pretty much the same. In American "oval racing" you're mainly "turning left" since the race is run counterclockwise (Europe is the opposite).

The track is slightly banked in the corners. The "right side" tires see more wear and tear because centrifugal force is always pushing the car to the right. You're turning one direction all the time [while] the rear wheels drive [the] car; you run stagger. If you were to just put the car "square" on the racetrack, you'd have the rear wheel both driving and trying to turn the car all the time. So you do "fine" adjustments with your cross weights, and you generally do it with a front end of the car. What the weight jacker does is make changes to the cross weights on the car while you're out there driving. In simple terms, thirty pounds of right front weight means push or understeer. Thirty pounds of left front weight means loose or oversteer. These driver adjustments have always been there because your fuel load changes the balance of the car. From a full tank to an empty tank, the balance will change in the car. Tire wear: the front tire could be wearing quicker than the rear tires, or vice versa.

So as you go through the race, track conditions change. Drivers need to be able to adjust their cars for just pure controllability and safety. During qualifying you might use the weight jacker on the straightaway. Say the wind is blowing; you're going down the front straightaway, you have a headwind, and when you go down the back straightaway you have a tailwind. On a headwind and turning left I'll go into the corner; all of a sudden I'll have the wind pushing a car to the left. You would feel it immediately; it would help the car to turn left. As I go down the next straightaway and into the corner the situation is reversed, so I'd quickly adjust for both ends.

Racing at those speeds, you have to anticipate everything. You actually have very good visibility, but you have to look up and [see] very far in front of you. You're traveling a football field per second. Another way of putting it: you need to be looking "around" corners. In order to "sneak a peek" at your gauges, let's say at Indy, you come out of turn two and take a look down the back straightaway; if there's nothing there—no cars, no nothing—then you start looking at your gauges and make sure that your instruments are in their "operating area." Then you look up right away because you've traveled quite a long way. Very seldom do you have time to look at your gauges when you're in the short chutes. There was a time when as you came through turn one you could straighten out the car and prepare for turn two. Now

Autograph day—the day before the 500.

that we're running 220 or 230 there is no more short chute. It's just one continuous turn.

I love the cars. I'll tell you, those years with Roger, those were the best cars that I've ever driven. My favorite car was the PC 27. What was so "tricky" about this IndyCar was the gearbox. The gearbox had oil veins from the engine oil that ran through the casing of the gearbox that would heat up when the engine heated up. The gearbox runs a different weight oil than that of the engine. When you're warming up the engine there had never been a system that warms up the gear oil. If you get everything warm then your tolerances can be a lot closer. Roger built that. The gear clusters on all other cars are geared differently from the engine rpm. They had something we call a "drop gear." Imagine the crank at its rpm and it bolts on to a gear that drops down into a different gear that holds your gear cluster. That holds first gear, second gear, etc. The gear cluster sits above the crank. What Roger did on this car is he made it a direct drive. They got rid of this drop gear. They did this to lower the center of gravity in the car. Get the center of gravity, the weight of the engine, and everything as low as possible and then there's less "roll" in the car. It took me three or four days of testing before I finally figured out how to shift the thing. But then I finally learned I could shift that thing with two fingers.

I love developing race cars. Now, everybody has to use the Xtrac gearbox; everybody has either a Dallara or a G-Force chassis. There's not as much of the individual development as there used to be.

Now the challenge is to see if you can get more speed out of equipment that everyone else has. Everything is superbly engineered. I raced the G-Force for a couple years and I raced the Dallara for a couple years. They're very predictable and they give the driver a lot of confidence.

## 23
# BOBBY RAHAL

*Won Indy 1986*
*Date of interview: March 2004 (This telephone interview took place as Rahal was in Florida for his Hall of Fame induction in 2004.)*

**YOU DON'T** get inducted into the Hall of Fame very often. This is the Sebring Twelve-Hour, the Grand Prix of Endurance. In Florida I won that race in 1987. Me, Mario Andretti, Sterling Moss, and a couple of other guys are being inducted this weekend.

**ON HIS REPUTATION AS A "NEW TYPE OF DRIVER":**

Maybe it's my glasses or the fact that I'm a college graduate (if I'm not the first, I'm among the first). There's a generational change there. Rick Mears got there a couple of years before me; he came up primarily through road racing just as I came up through road racing. That probably prompted the "new breed" nomenclature.

I think one of the problems in racing people are always asking is, [is] it the car? Is it the driver? You can't define it that narrowly. I think *one* of the main components of winning is being prepared. When an opportunity presents itself you take advantage of it. I don't know if you can ever win without everything absolutely clicking that day. You certainly can't rely on luck. I don't believe in luck. Everybody's got to do

their job. Success is based on the guy that's bolted the engine together, the guy who made the tires, the guy who mounted the tires, the guys doing the pit stop, the driver's got to do his job . . . there's so much that goes into it. I think that's why so few people win all the time. It's not based on money. You've got to have an appropriate amount of money, but money in and of itself doesn't do anything. You've got to have clear direction. You've got to have great leadership on all fronts—the owner, the driver, the crew chief, the chief mechanic, the engineering guys— and that clearly wasn't the case twenty-five years ago. One needs an absolute commitment to be the best. All too often you try to explain it, and you can't, other than to say that it all has to work together.

## ON PICKING UP BUDDY RICE AFTER HE HAD BEEN DROPPED BY ANOTHER TEAM:

I think it was due to a combination of things—the right car, the right engineering leadership, and Buddy is a great driver. He also

Like many winning drivers, Bobby continued in racing as a team owner. Here he gives a pre-race bit of encouragement to his driver Danica Patrick.

Rahal's pre-race steadying of his young driver Danica Patrick.

had the backing and the wherewithal (I don't mean financial; I mean psychological backing and leadership within the team from me on down). He didn't have that last year. Eddie (Cheever) is known to be a difficult guy to work for, if you're not delivering the goods. Eddie's way of solving that problem—historically—probably doesn't work that well. He's an emotional guy, and that's fine. Whatever was going on, he didn't click with Buddy. Some drivers need more motivation than others; guys are self-motivating. Some people are very smart behind the wheel and some have to be reminded to be smart. I think with Buddy, he realized that this was a great opportunity to come to our team based on the record of success that we've had. It was his chance to, sort of, silence the critics, his chance to show everybody what he could go out and do. He certainly succeeded at that!

## ON BEING APPROACHED BY UNEMPLOYED DRIVERS AFTER KENNY BRACK'S ACCIDENT:

I think you have to look at, "what are you asking this person to do?" What is *he* looking for? A year-long deal? If a guy brings with him a couple of years of sponsorship you can maybe take more of a risk, you can be more patient, but if you're looking for a guy that can run up front straight away, and if he's a little bit flexible in terms of longevity within the team, all of this scares you; so you try to see what the situation is, and you put the right peg with the right hole. Buddy knew that if he did the job he could leverage that into a full-time ride, maybe with us or with somebody else. Certainly he did a lot toward that. In our case, with Kenny, we intend—and we assume—he'll be back after he's recovered. We had to find a person who would fit within that demand. After the job he did at Homestead in Florida I can tell you that we're doing everything we can to find Buddy a full-time deal. We're definitely looking to run a second car full-time. We are doing so this year at Motegi and Indianapolis with Roger Yasukawa. It's very hard to be competitive if you're a single-car team, particularly in this day and age where testing is restricted so much. Two cars is twice the information. When it was "unrestricted testing," that wasn't a handicap for a single-car team, but now—especially if it's down to six or eight days—the more cars you have on the track, [the] more information you're going to get. They restrict you to the number of days you can test. At the moment there's *no* testing. You can do wind tunnel testing all you want, but with spec parts there's limited benefits to that.

We're starting this year with a new chassis. Last year we had Dallara. This year we have G-Force. That's a little bit of a negative because our data—our "book of knowledge" from the Dallara last year—we wouldn't want to use that. We were generally uncompetitive last year, so you go with the car that you think has the most potential. It's basically common sense. I think oftentimes you can over-engineer something and convince yourself that this is the right way to go—and of course it isn't. You complicate it. Yeah, there's science to it, but after all, if it's a certain way, go that way. Having a chassis for several

years, particularly if you've been successful with it, provides you with a backlog of information you can use.

It can be a costly decision; financially, it's also very risky. What if the car you pick isn't so good? We've got all of our Dallaras from last year, and we're going to keep them. If all of a sudden they appear to be far and away the best chassis, we're going to have to repair them and jump back into them halfway through the year. It's all about being prepared!

## Q: YOUR MOST VIVID INDY MOMENT?

Winning it! Winning it for [owner] Jim Trueman, who died a week later.

It was pretty unbelievable. It was an unbelievable month of May for me personally and for us as a team. It was a bittersweet moment, but nonetheless something you could never forget.

# 24
# BUDDY RICE

*Won Indy 2004*
*Date of interview: May 2015 (Carb Day 2015)*

Buddy Rice at the 2015 race lending a hand as a spotter for one of Ed Carpenter and Sarah Fisher's drivers.

The one thing I remember during the race [was that] we had stalled earlier in the race; they knew "weather" was coming, they just didn't know when. Once Scott Roembke, who was the team manager for Rahal (as well as he did all my strategy and "sat on my box"), said, "It's time to get going. The weather is coming. We gotta go! This is going to be a race to the end!"

That was a big thing. We made a desperate move toward the "big charge."

We were doing what a lot of guys were doing—just monitoring fuel. We were easily [*laughs*] . . . I say easily . . . running the car easy, just trying to save everything for the end, to see where we were going to be. But once they knew that the race was going to be shortened because of the weather and it wasn't going to make it, we had to get going and boogie. So for the next pit stop or two it was just flat-out—all the way . . . just started going for it. That was the big thing; the car was good enough that we were able to charge to the front.

When Roembke says it's time to get going . . . .

One, we stopped saving fuel. We knew we were probably going to have one more stop (regardless), maybe two; you just have to throw that out the window. Everybody is going to be similar at that point. Everybody is monitoring the weather. So at what time do you want

At the time that the photo was taken, Buddy raced for Cheever Racing / Red Bull. Buddy told a story of once getting a last-minute opportunity to qualify for a race in Riverside, California. He'd never as much as driven the track previously, and weather precluded any practice time at all. He only knew the racecourse from a video game he'd played. The simulation proved to be spot-on for the real thing—he qualified easily the first time out.

to go for it? The big thing with us is we got more risky with the car because you need to start driving to the front. You don't just sit there cruising, being patient, not taking any big risks. When I got "the call," I don't remember all of it. I was sitting sixth, seventh, eighth—somewhere around there.

We got rolling and took off. Even though you're passing cars, you have no idea where you're at. You're just focused on what you're doing.

When I started getting close, and knew I was up to the front and learned where I sat position wise and what was going on, then I knew that we had something and had to keep running hard. We made some changes in the car, started taking bigger risks, and then the miles-an-hour started coming out of it. That's what we had to do.

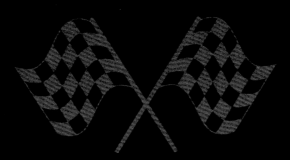

## 25

# A. J. FOYT

*Won Indy 1961, 1964, 1967, 1977*
*Date of interview: October 2015*

It seems that A. J. is always holding a stopwatch.

**Q: YOU DROVE ALL KINDS OF RACE CARS. YOU WON INDY TWICE WITH FRONT-ENGINE ROADSTERS AND YOU WON TWICE WITH REAR-ENGINE CARS. WAS THERE MUCH OF A LEARNING CURVE SWITCHING FROM ONE TYPE TO ANOTHER?**

They were two different types of cars to drive, but there was no learning curve; you either adapted to them or you didn't. It didn't bother me which one I was in, I just adapted to it. It was easy to me, the rear engine cars were a lot easier to drive than the roadster.

Just one of Foyt's championship rings.

**Q: AS AN OWNER, YOU WON THE 1999 INDY 500 WITH KENNY BRACK—A GUY FROM SWEDEN. YOU CURRENTLY HAVE A GREAT DRIVER IN TAKUMA SATO, WHO IS FROM TOKYO. SOME MIGHT SAY THAT'S AN ODD FIT FOR THE GUY FROM TEXAS. WHAT QUALITIES ARE YOU LOOKING FOR WHEN A YOUNG DRIVER COMES ASKING FOR A JOB?**

I look back at what he raced before and how well he did, and if he was a charger. I never did test many drivers, but when I do—in cases where

I don't know that much about them—I like to see how well they adapt to the IndyCar, and when we made changes, to see what he thought it did—the feedback. That's what I look at, so when you make a change, what did he feel—(and forget the computer).

That's A. J. high atop his team's "transporter" at Chicagoland Speedway. It's likely he's either testing a new prospect, sizing up the competition, or watching his driver turn laps. His ever-patient assistant Ann Fornoro is up there as well.

**Q: ON HIRING KENNY BRACK AND TAKUMA SATO . . .**

Well, Kenny I watched when he drove for Rick Galles, and he always ran real hard, but I don't think they could control him 'cause he ran so fast—and he'd run hard till he crashed. But one thing I liked about him was that when I talked to him, he'd listen. If I said, "Take it easy," or "Back off a little bit," he would. I don't think he respected the other people he drove for because you'd maybe see him out front but then he crashed toward the end. As far as Takuma, I didn't know that much about him, but we knew he was a good road racer since

A. J. Foyt IV, known as Anthony, driving for his grandfather.

Three generations with the same name—A. J., Tony, and Anthony.

he ran pretty good in Formula One, and with our series turning more toward road racing, that's what we were looking for. Plus he showed he could run decent on the ovals, and he was a charger. Now after three years, I think we've got him settled down. Before, if he wasn't winning, he'd run hard enough till he crashed, but like Larry [A. J.'s son who is president of A. J. Foyt Enterprises, Inc.] and I said, "If you've got a fourth- or fifth-place car, that's where we'd like to finish."

## Q: ON THE HULMAN AND GEORGE FAMILIES . . .

When I first went to the Midwest, we got to know Mary and Elmer [George], and Mr. and Mrs. Hulman and became real close. We spent a lot of Christmases together. The George family would] come down here for Christmas. Our kids and their kids were close together in age, so they grew up knowing each other, and even though they weren't kin to us, it was like we were family. It was just a close relationship and it still is.

## Q: A GOLDEN INDY MOMENT?

The first time I ever qualified for the Indianapolis 500 was a golden day for me, and then to be lucky enough to win it on top of that! My big goal in life was to be good enough to someday qualify for the Indy 500, so that stands out to me—the first year I qualified for the race in 1958. It was a big thrill.

Coyote Racing.

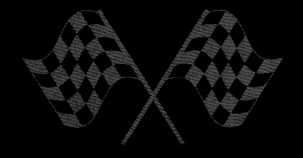

## 26
# KENNY BRACK

*Won Indy 1999*
*Date of interview February 2004*

**I SPOKE WITH KENNY BRACK AFTER HIS 2003 CRASH, ONE OF THE WORST IN RACING HISTORY, ABOUT HIS REHAB AND ABOUT RACING IN GENERAL.**

I feel good. Now, it's the weekend. [On] weekends you get two days to catch your breath; on Monday you start again. It's not a daily progress, I'd say, but it's a weekly progress. I'm in Ohio now, but I've been all over the place. I started this thing in Dallas at Parkland Hospital; the doctors there did a fantastic job and then transferred [me] to Methodist/Indy, where Dr. Scheid did a fantastic job. Then I transferred to Columbus.

**ON GROWING UP AND LEARNING TO RACE IN SWEDEN:**

Värmland is very much like Ohio. In Sweden we have a lot of different counties, or states you might call them. Not like the United States obviously, where each state has different laws and things, but it's still kind of split up into twenty or twenty-five different "regional areas"; my area was Värmland, which contains a lot of small towns. Värmland is the actual territory name, or state name. Glava is actually my home village, a little village right in the middle of nowhere. I guess you

could call it a "nature reservation." That's perhaps contrary to being in the business of racing, but it's rural and we used to play around with cars in the wintertime. Driving on the ice . . . when the lakes froze we would plow a little circuit on the ice. We had normal road cars, but we slid around. That's how this thing started. My dad used to teach me how to drive.

You know the World Rally Championship? They go all over the world, but in Sweden it's the Swedish Rally. They ran special events just across the lake from where I lived, so my dad and I would drive across the lake [laughs] when it was frozen, of course, and look at the rally cars that were driving by. When we did this—I was six or seven—my dad let me drive the car. I sat in his lap and just steered for the first couple of years (I couldn't reach the pedals). You have a lot of freedom to do whatever you want on the ice; you can slide, you can spin, you can do whatever you want without getting in trouble . . . usually. There's a lot of "run-off area" there, you know? [laughs] That's how I started. Soon I got to drive on my own. When I was about ten, my dad got me my first car to drive on my own on the ice.

We also had lots of small old roads. These roads ended in the middle of the forest. They had been used by trucks carrying timber out of the forest.

These roads, they went into the forest, and they stopped. If you went in there and checked, there were no cars there. And [so you'd] drive (slowly) out, and then you can drive like a maniac . . . in again. You knew that nobody would come from the other direction. It was "solo" (we weren't racing against anybody), but sometimes my friends were along with me. Then, a friend, just a neighbor at the time that I didn't know so well, he owned a bed-and-breakfast (tourists come from Germany and everywhere in the summertime because it's a nature preserve). He was into racing; he had raced himself. I started for him one summer; he saw that I could drive, and so he bought me a go-kart. I worked to pay him back. But he really helped me. That's how I got into racing.

There were organized events, but usually I practiced. In front of the bed-and-breakfast there was a big space that had asphalt area; that was the only asphalt place in the whole village at the time. There, I ran the go-kart. He helped me. We set up cones. From that it just evolved.

I got a license. We ran a few races, but it was summertime; there were other interests that you have as kid.

## WHEN HE STARTED IN THE "BIG LEAGUES":

When I started racing bigger cars I started paying attention to US drivers and racing—A. J. Foyt, Bobby Rahal, Emerson Fittipaldi.

It was amazing that I drove for Foyt for a couple of years. I actually won the 500 driving for Foyt. I got to know him—he's a crazy guy! [laughs] He's a different breed. He's an original person. It was great to hook up with him. He obviously knows a lot about oval racing, and that's what we were doing at the time. I was hooking up with one of the most colorful persons in racing. A. J. gave me a lot of coverage in media. It was great for me. We won a lot of races together—championships—and the Indy 500. That was good. Since then I raced for Bobby Rahal and Chip Ganassi; those are also great guys. Bobby was a class act as a team owner and very good as a driver. Chip has got a really great team. A. J. though . . . .

I drove for Galles Racing when I first came to United States. He gave me the chance to drive Indy cars. But Foyt was the team where I came into my own. I think what was good about driving for him was that he opened a lot of doors because of who he was. It was great for me, being associated with him. He knew how a car should be set up. That's something that's very difficult when you're starting something new—*to get a feel for what a good car is.* Today's racing has gotten into . . . it's always been very technical, but today everything is run by computers. When things get run by computers, everything is very exact. If you know how to use computers you can be a lot more exact than otherwise. If you think about the whole racing scene in the '60s, '70s, and '80s where computers were playing a minor part in racing, you had to do your setups from your fingertips rather than the computer. A. J.'s got an extreme feel for what the car needs in that situation. I can see why he won a lot of races during that era, because he's got an extreme feel to set the car up. He's got that "hunch" deep down—in his stomach or whatever—to feel what the car needs. That's [a] very delicate task, and I think he was pretty much the best at that.

If you took away the computers, A. J. would still put the best setup on the car. He's very, very savvy.

When I drove for A. J. it was in '98 and we were leading the Indy 500 when we ran out of fuel. That probably cost us the victory that year. I don't think it was the computer's fault. There was probably a guy running the computer who made a mistake or whatever. A. J., [though], he had to blame it on something, so the computer took the blame. He smashed the computer against the pavement at the timing stand.

With the teams I've driven for, there is not only one recipe to success. Foyt is very hands-on. He runs the thing; it's a one-man show. He directs all the time. Chip, for example, he's got a big team. He delegates, like a big company does. I would say that's a very "engineering-driven" race team. They look for the future in technology; they always want to be on the cutting edge of technology, to help them be more and more successful. Penske, I would assume, is the same, but I can't really say because I never raced for them. Rahal has more of a conventional, modern race team. They utilize the technology that's there for them, but they do try to develop some new technology. But they're not as "driven" in that area as the Ganassi team, or whatever. Rahal runs his team where he does not have to be involved in every detail. He delegates, trying to put the right people in the right place, like a big company. Definitely a difference in each team, but like I said, there's not just one recipe for success.

In Foyt's team the strength is in Foyt himself—his control and everything. In Ganassi's situation the engineering-driven side to it is Chip's "mission." He's got all those great people around there, running it. With Rahal, its strongest asset is that "team feeling" and getting the most out of what they have.

For a driver, it's all about the relationship that you manage to develop. I don't think it's the same for two different drivers. It's like any other place you might work. You might get along great with the people there and you do a really good job, but the next guy who might be as skilled as you are may not find himself at home at all. It's the same thing. In racing, it's very cutting edge. At a "normal" company, usually a lot of time passes before you can measure your success—good, not so good, or whatever. But in racing, it's so *critical!* Every part of a race team has to gel together perfectly; otherwise you'll see it in the results straight away. And it's not like we're running "close of books" every

year or every quarter. We see the results of our work every Sunday. It's extremely visible. Any race team runs very efficiently, I think, compared to a normal company. Everybody on the race team, they're there because they want to be there. They are there because they're exceptional individuals. If you get these people together, and they like each other, they develop this special bond. That group is hard to beat.

In a race team nobody is more important than another person. Everything is needed to win; it's not just about a great driver. It's not just about a great engine. From the guy that sweeps the floor—their effort, their input, everything is as important. If one link is missing, it'll show in the results.

I've raced for a lot of different engine companies. I raced for Ford, I raced for Oldsmobile, I raced for Toyota, and now I race for Honda. I think Honda, if you look backward at the history of motor racing, they have had tremendous success. They've won in every category they "set their hand to," so to speak. Lots of racing—Formula One, CART—lots of race wins and lots of championships. They get the most out of the technology and know how to use it. That's for sure. They're very skilled.

I had a very good relationship with Ford; we developed their engine in CART in 2000 and 2001. Toyota as well—I won races with them in 2002 in CART. There's very little difference among the engine manufacturers, because they're all very good. They're quite different, actually. They look at technology in different ways. One engine might be very good in "top-end power," which suits some of the tracks that we go to. Another engine might not have as much horsepower but might have better drivability, or more torque, or better horsepower lower down in the "rev register," and that engine might suit some of the tracks we go to better. It's all give and take. In the end you've got to have the package that "on average" is the best.

You're not going to be the best at every track. It's very difficult, especially in American racing where the rules are so tight. The end product that anybody makes will come out similarly. You might have 700 horsepower in one engine and 690 in another; it's not going to be a huge difference.

I think a driver that runs at Indy or any other track feels that they have a really strong team of safety people. Racing is dangerous, and all the drivers know it. If and when something happens we all know that we're going to get professional help. That's a very comforting feeling when you think about that. You know that you're not going to get somebody that might twist you around and make things worse. All these guys, they've been around for such a long time and they've seen so many things. You know that you're going to get the best possible treatment from the moment they get to you. Most of the time with car development and track development, the risks have been reduced. I think that the safety personnel have been a part of that. They've seen so many situations that their input has been very important:

How to make the seats safer, how to get an injured driver out of the car in the best way, transportation to the medical center, the helmet, etc. The helmets today are pretty sophisticated; they have a little plastic bag in the middle of them, and . . . you don't have to pull them off anymore. They attach an air-pressurized hose to that hose on the helmet, and it's like blowing up a balloon between the helmet and the driver's head. It pushes the helmet off the driver's head instead of somebody having to pull it off. Obviously when you pull it puts pressure on the neck; if you have a neck injury, that could be dangerous. So there's a lot of effort that's going into safety. Especially for the Indianapolis Motor Speedway's new safety barrier. That's supposed to be better so that when you *hit*, you're not going to hit as hard. The safety crews around the track include doctors, who assist not only people who have gotten injured, but also if you've got a bad cold you can go there and get treatment. They take really good care of everybody in the racing business.

## ON HIS BIG CRASH:

You can't afford to kill all drivers; racing has to be a civilized sport. Although I think that this is the closest thing to a "gladiator" sport—but it's still safe, you know? How should I put it? Racing is a high intensity sport with elements of danger—and that's what it is. I don't think you have any sport that's quite like that; look at basketball or hockey—there's nothing quite like racing. It's not like heavyweight boxing, where the intention is to knock your opponent unconscious. Racing has a dangerous element in it, but the intention is to win but nobody gets hurt. In boxing you can't win unless the other guy gets hurt; racing is not like that. The element of risk is there a little bit. The intensity, high speeds, excitement . . . there's nothing quite like racing out there.

People may not see this, but everything we do in racing, the public benefits. Innovations will show up in passenger cars down the road. Every crash you see in racing . . . they're logged and all this information is used to make passenger cars safer. Engine developments can be used to get more power to have less consumption of fuel. There's tire technology. There are a lot of different safety devices that you'll eventually see in passenger cars maybe ten years down the road. The general public at the same time benefits from this sport. You can't really say that about many sports these days.

My crash in Texas was apparently over two hundred g's. It's the highest measured g-force in any crash. Yeah, it was a bad thing, but I'm still here and can talk about it. If you look at it, the car really did its job; it absorbed as much energy as it could without the cockpit breaking apart. I got seriously hurt, but I guess you would if you hit something that hard at 220 miles an hour. I have no problem talking about it. I just know that that's what happened.

Brack performs a high-volume set for the raucous Carb Day fans. His group is Kenny Brack and the Subwoofers.

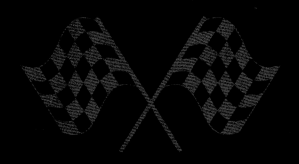

## 27
# JACQUES VILLENEUVE

*Won Indy 1995*
*Date of interview: May 2004*

Here, Jacques has returned to Indy for the 2014 race after a very successful career in Formula One.

## ON WINNING IN '95 AFTER OVERCOMING A TWO-LAP PENALTY:

Yeah, there was a new rule that year, about the pits being closed when there was a yellow flag . . . Just before the yellow flag came out, I'd been running fourth, I think. The leaders had come in the pits, and I ended up being the leader. But it was just about the same time the yellow flag came out, and nobody told me. Nobody on the team noticed this, and nobody told me. But every time I got to the pace car, the guy wasn't waving me to get behind him. I kept passing him, because normally the leader was to catch the pace car and then they would open the pits. That happened twice, and then they took those two laps away from us.

## Q: THERE WAS A SIMILAR INFRACTION BY SCOTT GOODYEAR LATER IN THE RACE, RESULTING IN A ONE-LAP PENALTY. DO YOU FEEL THAT THE OFFICIALS TREATED YOU BOTH FAIRLY?

What he did . . . we were all behind the pace car, with ten laps to go, and on the lap of the restart he just started too early and overtook the pace car to start the race. It was different; for his penalty they gave him a "stop and go," and he didn't stop, so they stopped counting his lap.

Of course the officials try to be fair. Definitely! The fact that we passed the pace car twice would create problems for other people; running out of fuel, it made it hard on other people as well. But taking two laps away is quite a big thing at the time, and we were quite annoyed as a team. But everyone decided to keep going. It's a long race, and things happen.

For the first one, we caught up to the leader at that time (and then there was a yellow flag) and overtook the leader in the pits. When we came out I was running ahead of that current leader, and I'd gained a lap. Ten laps later, there was another yellow flag. Because of that, I caught up with the pack before the pits opened. Then I was not far behind the leader again, and I just ended up racing and overtaking him on the racetrack. So I got another lap there. Then there was *another* yellow flag, so I caught up with the pack again, until the last yellow flag, when Scott Goodyear was leading and I was running second. The car was actually very, very good in the race. What was different from the way we usually race at Indy, we had to actually race aggressively, for the whole race (like "qualifying laps," which you normally don't do at Indy). But that was fun [*laughs*], because it meant getting sideways; it meant doing a lot of stuff you don't normally do at the Speedway.

### AS THE FIRST CANADIAN TO WIN THE RACE, IT'S IRONIC THAT HE WAS DUELING WITH ANOTHER CANADIAN, SCOTT GOODYEAR.

There was seemingly no way we could beat them. They were just so much quicker than us in that race.

Goodyear, our tire company at that time, had brought tires just for the race that had never been tested because we couldn't compete against them. The only way to beat him was to have him make a mistake. So, on the last laps behind the safety car, I kept going next to him and just getting him a little bit annoyed and pressured, and then he started the race too early and overtook the pace car, so that's what gave us the race.

### ON THE FACT THAT, FOR MANY YEARS, CART AND IRL EXISTED AS TWO COMPETING SANCTIONING BODIES:

For many, many years it damaged open-wheel racing in North America. It would take a huge amount—I don't know, ten or fifteen years—for it to become what it was. Nineteen ninety-five and '96, it was a great series. There was a lot of talent, and then the talent separated into two series and went away from North America as well. So now there's a lot less talent in these two series because of that.

And the IRL cars are extremely dangerous. They've had a lot of concussions, and that didn't happen as often when it was CART. I don't know if it's because you race on ovals all day. I've been talking with drivers who went from CART to IRL, people I know well. The cars seem to be too easy to drive, which makes very close racing, but it's not real.

The way the CART cars were, it was really, really difficult to get them to go fast. From speaking with people who then went to IRL, it's much easier to bring IndyCars up close to the limit, so even average drivers can actually drive flat-out, which means you have a bunch of cars driving close together in a pack. And sometimes there are drivers in the middle of it who probably should not *be* in the middle of it. That is probably what creates danger—that's all.

The Formula One car is a lot faster, a lot lighter; they did race the CART cars in Montreal on the same track as the Formula One. The Formula One car, with the driver, weighs 600 kilos.

### ON INDIANAPOLIS AWARENESS:

I was living in Europe and my father was racing in Formula One. Nobody even knew of "racing in North America"; basically, it didn't exist in Europe. That awareness happened during the 1990s, mainly.

My father, Gilles Villeneuve, moved to Europe and raced for Ferrari in Formula One. That's when I was eight. So I was too young when he was racing in North America to know "what was what." I only heard about the Indy 500 when my uncle raced there. His name is Jacques, as well. He crashed in "quali" and broke his legs. That was in the mid- or late '80s.

### ON BEING THE SON OF CHAMPION RACER GILLES VILLENEUVE:

What's wrong with that is that people know right away who you are. They don't wait a couple of years. Normally when you start racing you're at the back of the pack; you're young, and eventually you come to the front. Then, people start noticing you, even though you've been there for three years. When you already have a name, they know you from your first race; they expect you to win that first race, which obviously you don't.

It forces you to work under pressure, so you work faster. It's not a negative. If you survive it, I think it's a positive.

I started racing in North America. I did one year of Formula Three in Japan. A Canadian company wanted me to start racing for them in North America. We did the whole thing. By then, I was starting to know what the Indy 500 was, but I still didn't understand how big and important it was until I started racing in North America. That was in '93. That's when I really saw how big it was.

When you talk of Formula One, you're talking about the championship—the whole thing. There's no *one* track that stands out;

it's not like "Indy" or "Daytona." You could, maybe, say "Monaco" because it's downtown and it looks very special. . . . The most important race I've won, as a single race, was definitely the Indy 500.

## ON THE CROWD AT THE START OF THE INDY RACE:

You're aware of them, but the "tunnel problem" never disturbed me. I was always able to focus on the race. That was never a problem. But the race is long. Going around like that for three hours, it's not that you become dizzy, but you lose perspective a little bit.

The after-effect of winning the race is I just wanted to get home with my friends to have a good time together. I didn't want to spend the day doing interviews and all that; I'm totally different from other winners in that respect [*laughs*]. It was an important moment for me and the people close to me. So I just wanted to go—together—and share the moment, not to necessarily be in the media "eye" or the public "eye." That wasn't the reason for winning it.

I grew up in it, so it was never something I sat down and thought about. I just took it as it came, as a matter-of-fact thing. It's not something I'd been wanting or looking for; I just took it in my stride.

Now, in the winter I play a little ice hockey in Switzerland. I do a little ski racing, to keep the spirit . . . the racing spirit. Proper, licensed, ski racing. It's the same mentality. The difference in ski racing [is that] I'm not doing it professionally, so I don't owe anything to anyone. There's no pressure either—if I just want to crash, nobody else loses anything out of it. You don't have a whole team that loses out, or sponsors. There's no expectation; it's only for fun. The disappointment or the pride that you can get from it is as high as being in a race car. Mainly, because you know you're a race car driver and not a skier. If you achieve something in a car, that seems normal.

I think the sports are quite related—the mentality, the vision, studying the course—they are quite related. Now that the winter is over I'm back into my heavier physical training schedule, which has never been fun, but now it's especially harder since I'm not actually driving.

There's two things—there's stuff like running or bicycle or whatever, but that's really secondary. The other is cardio-related. Irwin, my "physio," my trainer . . . we've been working together since '96. We've developed together a machine [in] which we can control . . . humidity . . .

you name it. We do all the movements exactly the same as in a racer. It has my race seat, we work on the neck, everything muscle-wise. It's all computerized—very, very heavy-duty training. No one else has it; we did it ourselves. If you have something that will give you an advantage, you won't give it up, will you? [*laughs*]

## ON A. J. FOYT'S REPUTATION AS A MASTER OF PSYCHOLOGY AROUND THE RACETRACK:

The psychological side is what will give you the championship or the important races. That's what will make the difference. Because it's in "critical moments" that people forget who they are and it's their animal instinct that takes over. That's when they will start to make mistakes. And if they expect you to act in a way . . . or if they think you're stronger than them, for some reason or another, suddenly they will become weaker. So all the psychological things like that have a huge effect, on the important races.

Like I said before, the only reason that year that I won the Indy 500 is because I bugged Scott Goodyear on the two laps behind the pace car—that made him start the race too early and overtake the pace car. That was the only way I could win the race, and it worked.

It's like that in every sport; in sports, we're animals. It's true—the difference is [that] we are aware of what our actions can bring. That's different from the rest of the animals, but the more pressure you have, that's when your animal instinct comes out.

We're all predators.

Just look at the world.

Villeneuve with Tony Kanaan.

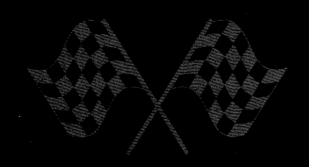

# 28

# GORDON JOHNCOCK

*Won Indy 1973 and 1982*
*Date of interview: February 2004*

Gordon Johncock's Borg-Warner image for his 1973 victory. It was a dark day for racing and especially for Gordy. His teammate Swede Savage was involved in a fiery wreck that would cost him his life several weeks later. A fire truck speeding to the site struck and killed his pit crew member Armando Teran. No celebration of the victory took place. A sweeter victory came nine years later, in 1982, when Johncock crossed the finish line about a fifth of a second ahead of Rick Mears.

WELL, THE first win in '73 was kind of bad in a way, you know?

It was rain-shortened. There were so many accidents. It was the race that everybody just wanted to get over with and forget about. We didn't have a victory banquet or anything. It was just a bad month of May. Period.

Even though it was rain-shortened, we led more laps than anybody else in the race. It wasn't a situation where the race was stopped and we happened to "be there" because of pit stops or something. We were very competitive for the whole race and had led the most laps, so it wasn't a "given" race to us, you know?

The second win in '82 kind of made up for it. I won the race by sixteen-hundredths of a second over Rick Mears. Probably the most exciting finish in Indianapolis history! There was one closer race ten years later, in '92, with Al Unser and Scott Goodyear. It was a little bit closer, but it wasn't as exciting, I don't think. You know, announcers have a lot to do with it. They make it exciting. My car was "going away," not handling well, and Rick Mears was catching me "a second to a lap." It was quite an exciting ending.

I knew that he was "coming on" 'cause my pit crew was telling me . . . on the "boards," so many seconds ahead, or so many seconds behind, and I could see him in my mirrors . . . coming. There was no such thing as a spotter back then, and I don't think they should have 'em now. I'm talking about a spotter up on a platform, who is telling you how to drive and who to pass, and all that stuff. Nowadays you have those spotters. I don't know. I don't think I'd want somebody telling me how to drive my racer, when to pass, this and that.

Rick had started on the pole. I started fifth. I was in middle of the second row. At one time during the race Rick almost lapped me. My car was working so badly in the first part of the race that if it hadn't been for a yellow, Rick probably would have lapped me. As the race went on we kept making adjustments. George Huenning, who took care of my car, the crew chief on it, kept making changes to the car and it kept getting better and better. He made changes to the "wings" on the car. The next thing we knew we were competitive and running with the leaders. If it had been several years later, after Rick

"mastered" the Speedway, I probably wouldn't have beat him. Rick wasn't exactly a newcomer, but he hadn't been there as long as I had, and he didn't know the right way to pass me. If he'd just done it right, he probably would have got by me.

## ON MASTERING THE SPEEDWAY:

Well, one thing I guess you've got to learn [is that on] corners two and four, you're going slower through those corners. You didn't need to "back off the throttle" so much, but nowadays they don't back off the throttle anyway. I think they're wide open all the time. You gotta learn how to get around the corners. You must know the condition the track's in, if you can run high [or] low. Know who you're running with, who you're trying to pass; there's different things like that.

## REGARDING TRAINING GROUND FOR AMERICAN KIDS:

The sport is so expensive; kids don't have the money to keep going.

Somebody needs to help these American drivers get rides and go on with their career. They don't have the money; the foreigners come in with their money, and "there goes the ride." That's what the car owner is looking for—money. That's what it's all about.

We didn't have much sponsorship back in the days; you've got to remember how long ago that was [*laughs*] when I started out. I ran "super-modifieds." Sometimes we'd run seven nights a week and twice on Sunday. I just got a lot of experience. We just ran, ran, ran . . . all the time. There's nothing better than experience. If you only run twelve or fifteen races a year you don't get much experience. I ran fifteen races a *week* in the super-modifieds and just graduated up the ladder. I drove for the same guy for ten years. Same way with the CHAMP cars; I had very few "different" rides. I was with the same owner most of the time; I was with Patrick Racing for a long time. Sure, there was prize money, but nothing like there is today. These guys today make more in one year than I did in my whole career. That's for sure. I know that.

But I like to be competitive, and I like to win races. That was what you thought about the most.

## ON WINNING INDY TWICE:

I guess, like a lot of us could say, "We should have won it more than that."

My biggest disappointment in Indy racing was I guess in 1977. I had a sixteen-second lead on Foyt with thirteen laps to go and my engine blew up. That took my third race and gave him his fourth race. So we should have been tied at three and three; that's one of those things that happen.

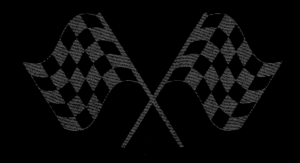

## 29
# GIL DE FERRAN

*Won Indy 2003*
*Date of interview: February 2004*

### HOW IS BRAZIL PRODUCING SUCH DRIVING TALENT?

I think that's a good question, but a hard one to answer. There's a great deal of passion for the sport over there. It's become somewhat traditional in our country. There has been a Brazilian at the forefront of motor racing for over three decades now—Emerson Fittipaldi in the early '70s, Nelson Piquet, Ayrton Senna . . . As a result of this continuous exposure, younger generations got inspired. They got inspired by these people, continuously inspired by all these great racers. Now a lot of young Brazilian kids want do that as their career.

It's a huge thing and I think it's been a good thing. The country is obviously not as prosperous as the United States. Motor sports in Brazil are more concentrated. There are certainly nowhere near the number of championships that there are in Europe or in North America.

This concentration has been a good thing for the young drivers. The more you face stiff competition the more you will develop. There's a lot of racing in Brazil, but once you reach a certain level, you want to leave the country. When I was racing there I became Brazilian Formula Ford Champion. That was the highest-level championship there was at the time. It's more developed now with the Formula Three championship

Victory salute from Gil de Ferran.

over there. But in my time if I wanted to "face the music," I had to go elsewhere, so I decided go to the UK, which is where everybody meets [*laughs*]. Most of the champions from around the world go over there to go racing. They try to showcase their talents to the Formula One team owners.

"Formula Indy" is a name that became traditional in Brazil. In the '80s when Emerson was instrumental in putting the series on TV, they didn't know what to call it. As far as they were concerned, they thought Indy cars looked like "Formula" cars. In the States you call them "open wheels," in England they're "single seaters," in Brazil we had called them "formula cars," and that obviously stemmed from Formula One. When they needed a name they came up with Formula Indy.

Three Brazilians, now with five Indy victories among them: Tony Kanaan has one, Gil has one, and Helio Castroneves has the rest.

Indy is known for producing the "highest of highs" and the "lowest of lows." It's a place of extremes. As I look back, qualifying was an *extremely* difficult day for me. The wind was howling. We went from having a very, very fast car on Friday night. I was very happy with everything, thinking that I had a chance for the pole. I went out the first thing Saturday morning with the windy conditions—slightly cooler—and I nearly crashed on my second turn. I thought, *Hmmmm, this is not much fun.* [*laughs*]

It was a very difficult day, I struggled all day. Probably a few too many "near misses." On it went. Eventually, I qualified tenth. Then, of course you have the *win*. It's the highest, I guess, you could have in motor sports.

[Gil beat out his teammate and fellow Brazilian Helio Castroneves, denying him a third victory. Helio joined Gil's victory celebration.]

Well, you know it's a funny situation. I've been in his shoes. In 2001 I finished nose-to-tail with him on his first victory. It's a difficult feeling. I remember what I felt—which is disappointment—not to have won the race, to have come so close to winning but not actually doing it. We were nose to tail all the way to the end. On the other hand, I was kind of happy. First of all, finishing second is no disgrace, and finishing a close second is even less of a disgrace. The third issue for that is that you are happy for the team . . . And I think . . . Helio and I, we have a genuine friendship, you know? So he won in 2001, and I won in 2003.

## ON HELIO "CLIMBING THE FENCE" WITH HIS WINNING TEAMMATE:

That was a funny moment. That's what *he* does. I've got nothing to do with that [*laughs*]. But the race finished. I did the parade lap, everything was great, the crowd was screaming, photographers everywhere. I

Interviews on the Bombardier walkway continued for hours.

Trackside interview just minutes after victory.

hadn't seen Helio. I knew he had finished second, but I hadn't seen him yet. So finally, he comes to me—after I got off the pace car—he comes to me and we talk. And when the crowd sees us, the crowd starts going mad. Everybody is just going totally nuts. So we both wave to the crowd—"Thankyouverymuch"—and then I think, *I've got to go now.*

At the same time, as Helio is walking away, and the crowd is still going crazy, I'm thinking: *This is weird.* So I had this idea, *Call Helio back.* So I had one of my guys, one of the guys working for our team, call Helio back. He came back and said: "What?" I said: "Look at this! We should go over there and . . . *do your thing* . . . together." Helio said, "Yeah, sure!" So we ran toward the fence together and did it. It was a cool moment—spontaneous—a happy time!

Gil with friend, fellow Brazilian, and Penske Racing teammate Helio Castroneves. After each of Helio's three Indy victories he "climbed the fence" Spider-Man–style to acknowledge the fans' support. Gil offered an impromptu invitation to Helio to join him in celebrating his own victory in the same way.

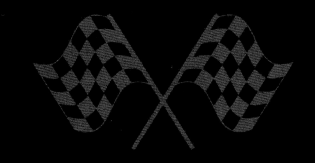

## 30
# EMERSON FITTIPALDI

*Won Indy 1989, 1993*
*Date of interview: November 2015*

**Q: IT SEEMS THAT YOU WERE AMONG THE FIRST OF THE CHAMPION DRIVERS TO STRESS GOOD EATING AS CENTRAL TO YOUR TRAINING REGIMEN. HOW DID YOU COME TO THAT NOTION?**

Like any athlete that wants to participate in sports . . . when I left Grand Prix racing and I came to race IndyCars I was already older than the average of the young drivers. I talked to people in my physical training program. I wanted to have better nutrition, and I found a very good guy in San Francisco called Gary Smith who had a lot of Asian influence—macrobiotic foods—what I call "high energy diet." I was already forty-three or forty-four years old and I mean, you remember I was racing against guys who were twenty-three or twenty-four. I had "double age" and I wanted to keep up—performing at the same level during the second part of the race—mainly on street circuits, 'cause on an oval the "physical" is not so demanding, it's more mental.

I started going through these new types of diets. First I went through a cleansing program (no sugar, no dairy products). I lost weight and got energy, but mentally I had a clear mind. Yes!

And now, these last ten years I'm even more into it than when I was driving. I believe it's very important—the physical part and the spiritual part. Many times in the Bible, God says: "We are the temple"—our body is the temple of God. We have to respect that—and we have to take care of our body. I'm from the Baptist Church (originally I was Catholic)—I was baptized in '96 after the crash in Michigan. I accepted Christ and then started to study the Bible. Even now, more so than ever, I believe we have to take care of ourselves; we have a sacred body from God—a temple.

### [FITTIPALDI'S FIERY CRASH AT MICHIGAN INTERNATIONAL SPEEDWAY AT AGE FIFTY VIRTUALLY ENDED HIS RACING CAREER.]

I think . . . I know with my experience . . . the physical training program is important, but my opinion now is to experience both—physical training and diet.

I started working out doing physical training when I was fourteen or sixteen years old—[practically] all my life. I believe now the nutrition is even more important than the physical training program. Both are very important, but eating properly is more important than a physical training program. We are what we eat.

### Q: FOR THE LAY PERSON, IF YOU COULD STEER THEM DOWN A CERTAIN PATH, WHAT SHOULD THEY BE EATING AND NOT EATING?

From my own experience I can advise, but it's difficult because each person has different needs. You know my opinion? You have to go to a nutritionist . . . to a doctor to examine really properly . . . an exam to know exactly what you should eat and not eat. From my own experience? Everything that's very natural . . . coming from the ground—no fats, no sugar, no dairy products because that's a different animal fat that you're putting in a body. No soft drinks, use teas, but no [artificial sweeteners], preservatives/conservatives on the food because that's very damaging for the future—the possibility of generating cancer. All the chemicals that exist in the supermarket are horrible.

### Q: ON A RACE DAY SUCH AS AT INDY (THAT BEGINS IN THE LATE MORNING), WHAT WOULD YOU LIKELY HAVE EATEN?

There are three important things. First, you should never have a very "heavy stomach" because if you need surgery because of a crash, it could damage the surgery. Always, before driving a race car you should be light . . . for emergency; that's the first concern.

Second concern? I would say something that has a lot of nutrition, but light. I would go for, like in the morning, a nice granola with rice milk or soy milk; that's normally what I would eat "before." I eat a lot of fish; people don't like it, but I eat fish in the morning . . . and a lot of seaweeds, I eat too, for protein.

For the future? The correct food for the future that is already happening now are the "natural foods"—you know, "organic." Organic is the future for eating properly.

The food industry, to supply "good taste," creates horrible things that damage the body.

I have a little boy, eight years old. He just won the national GoKart Championship. We conceived him macrobiotically. Myself and my wife. My wife does the same diet. He was conceived macrobiotically—and we don't give him medicine—he takes everything "natural." His name is Emerson; his nickname is Emmo. I've now transferred the name to Little Emerson. Everyone now knows him as Emmo . . . Emmo! It's amazing to see these kids race in Brazil—also in America [and] they race in Europe. They are so young; [it's] just so talented and incredible to watch that new generation.

### Q: THOUGHTS ABOUT INDY?

I have so many good memories from Indianapolis. When I joined Team Lotus in the beginning of the '70s I was having dinner with Colin Chapman. I was asking, "How was winning Indianapolis with Jim Clark?" I always created in the back of my mind the wish, the dream to race at Indianapolis. I never met Jim Clark; when I arrived in Europe (when I arrived in Italy), he had had . . . the crash the year before. He was a great idol for me, but I never had the opportunity . . . but . . . Indianapolis is part of my history—very important. I love Indianapolis.

The first time I came to Indianapolis, A. J. Foyt took me on a golf cart to show me the "four corners"—what should I do? what should I not do? It was 1974. I was testing for McLaren with Johnny Rutherford. Bobby Unser was also testing. I had just won the World Championship for McLaren; they asked if I would "test" at Indianapolis. A. J. was very good to me. He was giving advice: Turn 1-2-3-4 . . . if the car slides in the back, if you lose the back end, don't overcorrect because you can "T-bone" the wall. There's a lot of advice that A. J. gave to me. He was very good to me. He was extremely good to me.

The biggest moment? For me . . . was my first win, no doubt about it. It was a big challenge at Indianapolis to adapt myself—the cars, the oval track—[I needed to learn] everything from scratch.

## PLANS TO BE AT INDY IN 2016?

I'll be there. . . . It'll be 100 years; I'll have to be there, for sure.

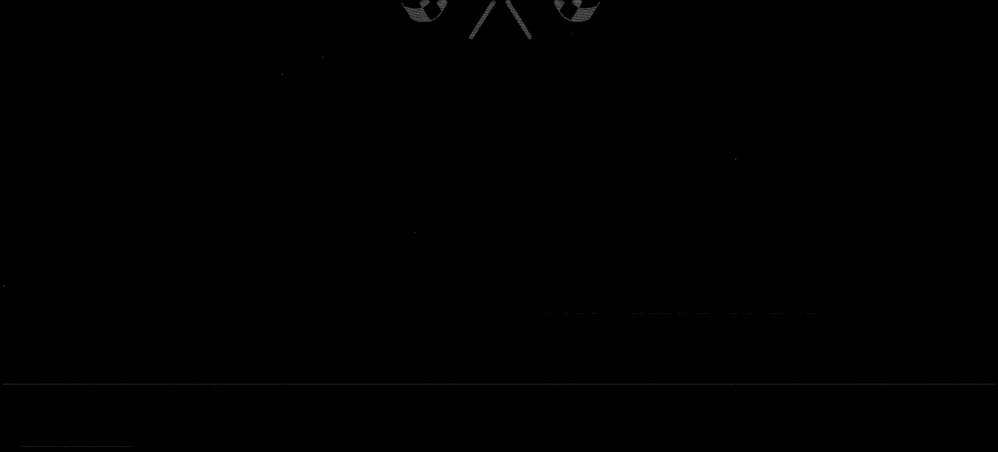

# 31
# HELIO CASTRONEVES

*Won Indy 2001, 2002, 2009*
*Date of interview: May 2015*

Helio is witty and relaxed. With three victories as of 2015, he is just one notch away from joining the elite group of drivers with four victories: A. J. Foyt, Rick Mears, and Al Unser.

Helio Castroneves, still wearing his helmet and HANS Device, scales the fence and pumps his fist for the appreciative fans.

Helio's Team Penske pit crew joins in the salute.

### ON WHICH CHAMPIONSHIP RING HE'S WEARING:

The last one. A whole drawer full eventually; that is the goal! [*laughs*]

### GOLDEN MOMENT:

I think you never forget your first. But you also remember very well your last. And the circumstances, especially [during] the last . . . I wasn't even thinking about Indianapolis. There were so many other issues that were happening . . .

When I came out there victorious it was an amazing, amazing accomplishment.

### ON GOING INTO THE "ZONE" BEFORE GETTING INTO THE RACER:

For me, when you put the visor down, it's like a shield when you go into a battle.

Watch out, man! [*laughs*] 'Cause I'm ready.

I'd been conducting an interview in the "drivers and VIPs only" section at the "Public Drivers Meeting" the day before race day. As the proceedings started I clearly felt out of place there. When Helio (pronounced "Ellio") and Sam Hornish strolled up, they paused to sign some promotional helmets that all the drivers were autographing. I had my camera with me and decided to request a "dual portrait" of the two Penkse teammates. I whispered in my best non-paparazzi voice; "Sam! and 'Ellio!" Neither budged an inch, but Helio said under his breath: "Salmonella?"

A meditative calm settles on Helio before he climbs into the cockpit. Like virtually all of the current drivers he is a superb athlete.

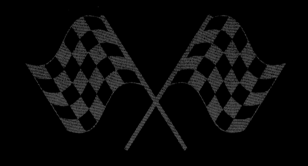

# 32
# TONY KANAAN

*Won Indy 2013*
*Date of interview: May 2015*

**AS A** driver, you don't come to participate in the Indy 500 not wanting to win the race, so there's no point of being here if you don't want to win the race.

That's probably the excuse I use: to get my other kid a trophy. [*laughs*]

But (confidentially) I want to win again, too.

## HIS GOLDEN MOMENT:

The restart. You know, I was sitting "second"? I remember I came on the radio and I said, "Jimmy . . ." Jimmy Vasser, team strategist, was on the radio with me. I said, "It's going to be all or nothing." And what I meant is I was going to go for it. I was

Kanaan is among the most popular drivers in the paddock. He's VERY proud of his nose. When he finally secured an Indy victory after about a dozen noble attempts, he relished the fact that he'd have his sculpted image added to the Borg-Warner trophy. Now, he quips that he's anxious to repeat the victory so that they can add another portrait to the opposite side and use the "noses" as handles.

going to take the lead. Maybe that was going to be the wrong move, because there were still three laps to go. When I made the move and a corner later, it went yellow, I knew I had made the right move.

That was it. That was "the moment." That moment created plenty of other moments, but that was the key.

Tony, in 2015, is racing for the sponsor NTT Data.

There's the famed Kanaan profile, on the rung right next to his pal Dario.

These fans wished it to be known that they are "family."

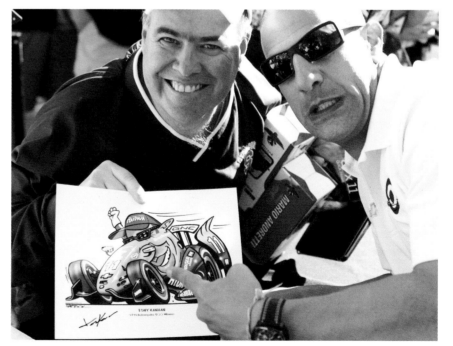

Autograph Day has fans bringing lots of things they wish their heroes to sign. Here, a fan has presented a cartoon of the Kanaan visage—as a race car. "TK" spun around to have the "real thing" in the photo as well.

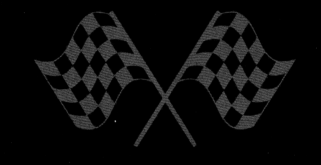

# 33
# SCOTT DIXON

*Won Indy 2008*
*Date of interview: May 2015*

## Q: WHAT MOMENT STANDS OUT WITH YOUR INDY VICTORY?

For me, I think it was probably the last restart. Getting a jump on Vitor [Meira].

I kinda went, slowed down, and went. I think it put him into a bit of a frenzy, which for me, at that point, was enough to start the momentum [and] to break the "tow" that enabled us to get to the front.

"Dixie," a New Zealand native, has a reputation as one of the most fit of the current drivers.

## ON PAST WINNERS:

It's a pretty small group when you look at it. It means a lot. Not just for me, but obviously for being part of a team. It's a *massive* team sport. It's never one person that wins this race. I think the "first order" when we won was to try to get back to the pits, to see my wife, the people, all the guys on the pit crew, strategy people, [and to] see Chip [Ganassi, team owner]!

You get *close* to winning races and . . . it hurts. Yet sometimes, talking to A. J. or Rick—guys who have won it multiple times—some of the ones they won, they didn't think they were going to win it that day. It just happens.

Dixon ready to race.

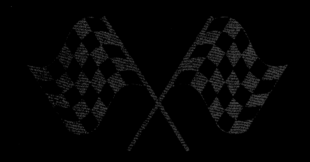

34

# DARIO FRANCHITTI

*Won Indy 2007, 2010, 2012*
*Date of interview: May 2015*

## ON HIS THREE WINS AT THE INDY 500:

The first one was 415 miles. It was a situation where Tony [Kanaan, teammate] and I were at alternate strategies, and I felt if the race had gone its distance then we would have had a hell of a fight because both of our cars were as good as each other's.

As it happened, my strategy is the one that came through. We had the car to win with. I'd driven it "from the back to the front" . . . after the rain delay . . . the timing was quite fortuitous for that. When rain finally came for good, yeah, it was a good feeling to finally "get one."

## ON MATCHING THE FEAT OF FELLOW SCOT AND RACING LEGEND, JIM CLARK:

I was very much aware of what Jim had done here. I obviously wanted to win the race for my own reasons, but I also wanted to win it, as much as anything, because he'd done it. I managed to finally make that happen.

Here at Indy you have to use all your racing skills. I think over the whole event it takes everything you've learned, in my opinion—the driving of the car, the setting up of the car, making the car "right." Also the mental battle that you face here . . . trying to keep it on an even keel [*laughs*]. Just doing your job. I think everybody on the team faces that battle, too. Experience counts here for a wee bit, to help you get out of those difficult days.

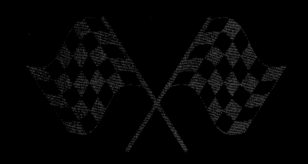

# 35
# JUAN PABLO MONTOYA

*Won Indy 2000, 2015*
*Date of interview: May 2015*

The engines are now quiet and Dad has been whisked away to "Victory Circle." The Montoya girls, Paulina (standing) and sister Manuela, indicate their approval.

Juan Pablo savoring his second time in Victory Circle—fifteen years after the first.

Solo shot at the Bricks with the "victory lap" car behind.

## ON THE IMPORTANCE OF PREPARATION:

Jimmy Vasser had a huge lead, and he was matching my speed at that time and I was worried that he was going to gain on me. I do remember that. You've got to prepare the car, work on it, make sure you have a good car.

See how it behaves. See how you're stacked against everybody else. See what you have to do to win it. It's the whole race.

You can work on it and be in the perfect position and on the last stop. You have a bad stop, and you've lost everything. You've got to be prepared for anything.

## ON BEING AMONG THE OTHER ELITE WINNERS:

To be honest with you?

I dunno. I always tell people, "The day I stop racing, I'll look back on what I've done and say, 'That was really cool.'"

Right now, I focus on what I need to do.

*Note: One day after this short interview, Montoya won the race for the second time. He seemed to savor the moment and included his young family in the celebrations.*

**Autograph Day 2015.** Still a one-time Indy winner in this picture.

In his 2000 winning image he is "Juan Montoya." His 2015 image will include his famous middle name.

How often are you likely to see Roger Penske wearing his cap backwards?

## 36
# RYAN HUNTER-REAY

*Won Indy 2014*
*Date of interview: May 2015*

### ON THE 2014 INDY RACE:

I think the pass into turn three was "one of the moments." When I pulled that off, it was a big moment. It disrupted the rhythm of the race a little bit, and Helio didn't pass me that lap, the whole next lap.

That threw the rhythm off a bit and kind of shook things up.

Helio got by me again after that, and I had to get by him again. There was still a lot of racing to do with three laps to go. I knew it was just between us at that point. That . . . that was a special moment, just knowing that I was very close to having (potentially) the biggest win of my career.

He's Tall, He's Smart, He's from Florida—2014 Indy 500 winner Ryan Hunter-Reay

Ryan Hunter-Reay's "private moment" at the Yard of Bricks.

Someone on the Andretti Autosport team has obviously been here before. All the caps are conveniently reversed.

Ryan's son Ryden is wearing a custom racing suit just like Dad's.

The Hunter-Reays. His wife, known professionally as Beccy Gordon, is the sister of former IndyCar driver Robby Gordon.

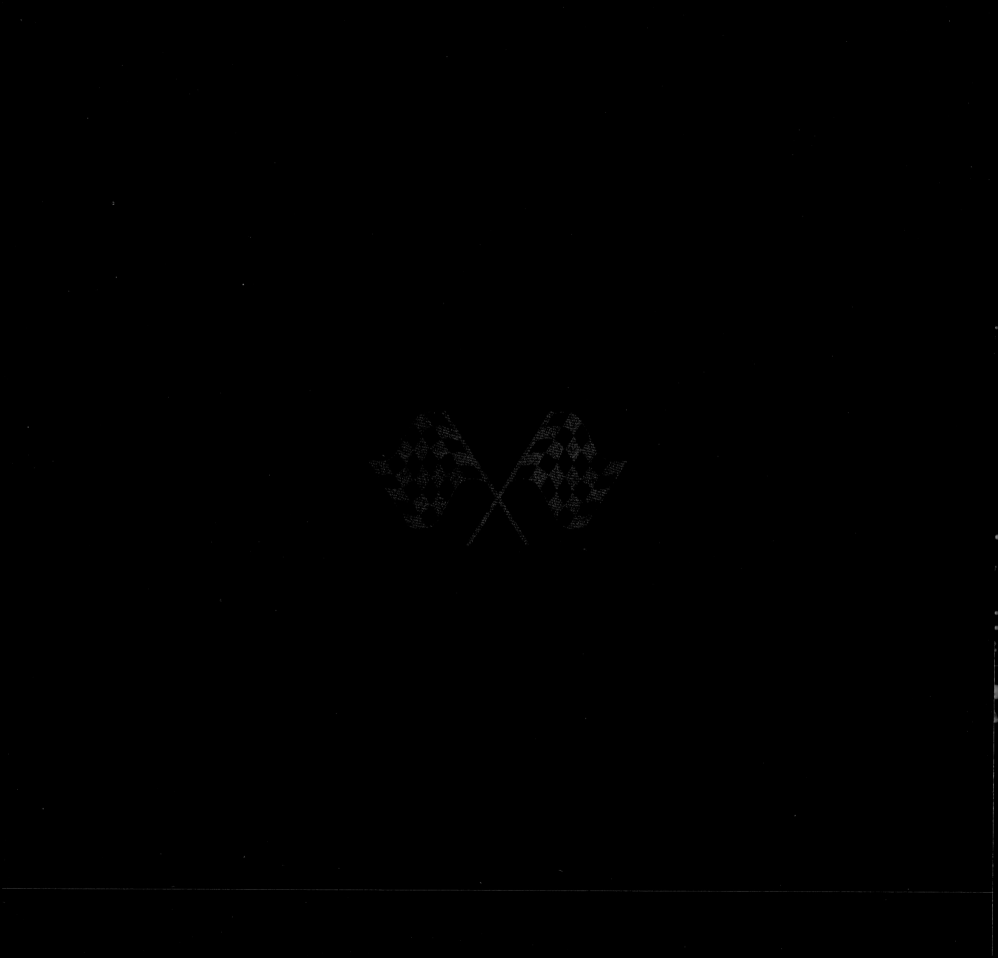

## 37
# SAM HORNISH JR.

*Won Indy 2006*
*Date of interview: December 2015*

**Q: I UNDERSTAND THAT OLD CARS KEEP YOU BUSY WHEN YOU'RE NOT RACING.**

I've got a '27 Essex. It's "titled" as a '34 because it had extensive damage. The body is a '27. That's all done. I've also got a '51 Mercury and a '67 Chevy pickup that I continue to work on.

**Q: ARE YOU GOING TO LOWER THE BACK END OF THE MERC AND BECOME WHAT DAVID LETTERMAN WOULD CALL A "LOW-RIDIN' PUNK"?**

[*laughs*] That's kinda where I was heading. I got the frame done—that was when I was still on the IndyCars. I got an engine for it and began doing some of the bodywork. Then I started running NASCAR and all those plans got put on a back burner because my residence is in Ohio—I have a shop there. But now, most of my time is spent in North Carolina. Being in NASCAR, they have a mandate. When all those projects got put on hold I started doing a bunch of "little ones" for the kids—go-karts and stuff like that.

**Q: YOU STARTED RACING IN GO-KARTS?**

These are more like "play" go-karts; A '60 Cadillac body that I bought on the Internet a long time ago. It uses electric parts that I got out of a

scooter to do the drive train for it. Then I've got the ones I'm working on now—a 1950 Jaguar "body configuration." I'm trying to use some Chinese "4-wheeler" parts—just so they've got four forward gears—and reverse. It's got a little alternator (for lights) on it, too. Stuff that's over the top and takes too long to complete, but I can do it. I've got a little workshop set up in one of the rooms of my basement. I can work on that kind of stuff down there.

## Q: ANY THOUGHT OF LAYING OUT A TRACK FOR THE KIDS?

I've got ninety acres in Ohio so hopefully at some point in time I'll be able to get back up there. These are more just "driveway" things, just to see if I can do it. I do it just for the fun of it. I'm doing a "two-off" (not just a "one-off") for both of my girls.

For my son, I've got a body that I bought a long time ago at a swap meet. It's about a mid-'60s IndyCar, a rear engine type, before they did "wings." It was a coin-op ride where you would put a quarter in it and "ride" up and down—like they used to have at the front of the grocery stores. Someone just took the body off of it, so I had that to start with.

Those three projects are what I'm trying to do all together. When I get 'em done the kids'll have their own little cars. I've got to hurry up. My oldest is eight. If I don't get it done . . . she'll be driving a real car by the time that happens.

## Q: I REMEMBER SPEAKING WITH SIMON MARSHALL, A TOP DESIGNER AT G-FORCE. HE CITED YOU SPECIFICALLY AS A DRIVER WHO UNDERSTANDS THE INS AND OUTS OF AERODYNAMICS.

The first car that I ran was a G-Force when I did my rookie orientation, not only for the IRL but for the Speedway. The first three races I ran was a G-Force—a "year old" one. That was back when they had it set on a three year . . . they were trying to contain the costs. The chassis manufacturers weren't allowed to do a full new chassis but every three years. So there would be updates. In 2000 they did one of those. So I was running like a 1998 car. The speed wasn't really there for it to compete with the new G-Forces and Dallaras, but with the team's sponsorship money, that's what we were working with at the time. We ran that car at Las Vegas and finished third with it. It was a day that was pretty hot and pretty slick. It wasn't about outright speed. It was

a little bit more about handling. We were able to start mid to the back of the pack and work our way up all day. That was PDM Race Team. That was the best finish that they'd had. We were able to make a deal with Dallara to get a new chassis to run at Indy. It was nice. It was a little bit more "in the ballpark" as far as speed and things like that.

We ran the G-Force twice more after that on short tracks and then ran the Dallara on "mile and a halfs" and bigger tracks.

## Q: SIMON MARSHALL SAID THAT YOU HAD A FEW TRICKS TO RUN IN "DIRTY AIR."

[laughs] There's a lot of things about the way that I ran IndyCars that other people didn't want to do. Like how close I felt that I was capable of running without making a mistake and wrecking somebody else. Really, for the IndyCar side of my career, I made a lot of things happen by how close I could run to other cars.

He [Simon Marshall] is right about where I would run. Getting close enough to 'em but just off 'em—just enough to one side or the other—keeping downforce on it to keep it "turning." Also, being on top of the adjustments inside the race car so I could run that close and get the car free enough when I needed to be, so that when I got that close to people—behind them—that when the dirty air became a factor the car didn't just automatically "go tight." It was tighter than what I was driving up until that point, but it wasn't like I was driving a "comfortable car." You get the dirty air, then it gets tight, so you've got to be willing enough to drive it "free" so that when you do get the dirty air the car then goes "neutral" and you don't lose all that front downforce, making the car unable to turn.

## Q: SOME DRIVERS SAY THAT THE "SHORT CHUTE" BETWEEN TURNS ONE AND TWO HAS BECOME, IN REALITY, JUST ONE LONG TURN. DO YOU FEEL THAT?

Indy is a lot different from most of the other tracks that we ran, as far as how you have to run it. It was about timing and knowing how close you needed to be to somebody through the corner so that you could get the proper draft in the straightaway to make a pass. There were a lot of times you could drive right up there on their gearbox, but if you didn't use the momentum to your advantage, there was no reason to do it because you were going to drive too close to them through one section or the other—be it one and two or three and four—that you wouldn't be able to stay flat on the throttle coming through the second corner and be able to make the pass. Indy for me was always the four corners because one and three are a lot different from two and four. When I was running there were definitely differences between the corners and adjustments you had inside the cockpit. You would, at times, make those adjustments in the short chute for the second part of that corner. You always want it a little tighter in one and three than you would want it in two and four.

The short chute is not a traditional place to pass, but it's a good place when other people make mistakes—say, going into the corner too close to somebody's gearbox in one and three and they'll be a little bit tight at the "exit" and they'll have to "lift." If you time it right you can take advantage of that.

## Q: YOU WON THE 500 RACING FOR TEAM PENSKE. WHAT CAN YOU SAY ABOUT ROGER?

Roger's done a lot for automobile racing at Indianapolis and he's reaped a lot of benefit from it as well. I think Roger would have given anything to just race. Roger was a driver himself—did a lot of sports car racing. But he was also successful as a businessman and it got to a point where he knew that to find what was going to benefit him most in life was to continue on the business route. I believe that a lot of the work he's done in business is so he can go to the racetrack every weekend. He loves it. He has been involved in sports car racing [Le Mans] and NASCAR, but I believe that open-wheel racing is his true love, and the Indianapolis 500 is his real passion.

For him, it's "How do we win?" I know that there have been times that his teams have won the Indy 500 and come close to winning the championship for the year. Roger would say: "I'd still rather win the 500 and finish second in the points championship than win the championship and come in second in the 500."

## Q: WHAT ABOUT YOUR INDYCAR SUCCESS BEFORE SIGNING WITH PENSKE?

I'd won two championships and we were on our way to winning the third and had an engine failure. We were leading the race when the engine expired. That was going to be my last race at Panther. We were pretty close to doing it.

A lot of people will say to me, "When you won the 500 in 2006, that must have been the best race you ran there." A lot of times I would say, "The best race I had at the Indianapolis Motor Speedway was a race I didn't win."

Two thousand three, in my mind . . . we had a car that was . . . we qualified, I think, 18th. We just weren't where we needed to be as far as speed goes. That was the first year that Honda and Toyota came in. We were behind, at Chevrolet.

We were running fifth with three laps to go when the engine blew up. The next Chevrolet was, I think, two laps down when we had that issue. I said, "Man, we did everything we could possibly do right." We were real close to winning three championships in a row at Panther.

## Q: YOU HAVE EARLY MEMORIES OF INDY?

My mom and dad for their first date went to an IndyCar race at the Milwaukee Mile. That's where my mom was from. My mom and dad went to the Indianapolis 500 when my mom was eight months pregnant with me in 1979 when Mears won his first 500.

I remember going to the 500 when . . . I remember being there and I know that I was there before I can remember being there. I was a fan of that race, growing up as I did, close to the 500. That's what it was all about.

About a year or two into my "go-karting days" when I started racing, I washed trucks for my parents after school on a weekday. "We're going to spend this money on racing so how can I contribute?"

I had a dream one day that I was washing trucks and this big motor home pulls up and it was Roger and Rick. Roger was trying to hire me and Rick was trying to talk me into going to work for Roger. I remember waking up and thinking, *Right, like that's ever going to happen.* It was just about ten years after that that I was in an office in Bloomfield Hills, Michigan, with Roger, talking about the opportunity to come to work for him. I thought, *This is surreal.*

I had a good thing going with Panther and I really appreciated everything that those guys gave me, but looking back, my favorite IndyCar driver, growing up, was Rick Mears. I was a fan of Danny Sullivan [and] a fan of Al Unser Jr. They are all "Penske guys." So, then, I'm a fan of Penske racing—or Team Penske, I should say. It turned into one of those things. When I was presented with the opportunity to go work for Roger, it was probably my best chance to win the Indianapolis 500. I could probably do it with Panther, but if I didn't win the 500, at least I will have had the opportunity of working with Roger.

We had a rough first couple of years running with him but I believe that they hired me to win them a championship and I went there looking to win the 500. In 2006 we were able to do both of those things. I feel like that was pretty neat. We were both able to fulfill the bargain that we set out to do.

## Q: DO YOU HAVE AN OPINION ABOUT MODIFYING THE DESIGN OF AN INDYCAR TO ADD PROTECTION AROUND THE DRIVER'S HEAD? IS THERE A "PURIST" POINT OF VIEW ON THIS STYLE OF RACING?

Things are a lot different from when I ran there. The "side pods" were about six inches inside from where the tires were. Now the side pods extend out past the tires, so if you're running on an oval or a road course, if you're completely side by side with somebody or at least halfway up on them, you can "touch" and nothing is going to happen really, as long as it's not too abrupt. Where, when I ran, you miss by half an inch and one car is generally "climbing up" over the top of the other one. When you want to talk about open wheel . . . open cockpit racing . . . some of that is already taken away with the design that they have now with the side pods extended out as far as they are. Now that being said, when we talk about a "cockpit" and things like that, I don't know what the right answer is. I don't know that a "bubble" . . . a bubble presents so many other things that you have to look at. Sometimes you remedy one problem and you come up with another one.

A bubble would be really hard as far as being able to keep it "visible" for a driver to see out. Especially at a place like Indianapolis. In NASCAR the windshield is basically flat [and works well with "tear-offs"]. Whereas with a bubble on an IndyCar [assuming curved surfaces] the tear-offs would tend to have air bubbles.

You need to have airflow inside the car. You'd basically be turning it into a "greenhouse" if you've got a bubble. That becomes a problem. You wouldn't think of an IndyCar as that hot, but the radiators are set on either side of the driver. There's heat that radiates from that. If you take away the airflow that keeps exchanging cooler air as you go along, it's going to become pretty hot inside the racer. For me, it would never

be a purist thing. If you put a bubble over the top, it's just a ridiculous idea. It's one of those things that's hard to make work. If there's a fire, that's an issue.

I really believe that at times there are things that you can do to make it safer. Everybody talks about "tethers" and things like that [strong cords or straps that keep broken pieces of a wrecked car from flying around]. Well, you can't tether everything on the car. Cars hit the wall hard, so one of the things that I continue to hear Rick Mears say, people ask him what he thinks needs to happen. He says, "The downforce needs to be taken away but the horsepower needs to stay." So basically, what it means is you have to slow down in the corners. Right now there's too many tracks where you're going flat-out . . . maybe 230 mph in the middle of the corner . . . Well, if you could go, say, 235 down the straightaway but have to slow down to 205 in the corner [in an accident], you'd ultimately hit the wall with less inertia and the pieces don't fly off the car as hard or go as far.

The "nose" is a hard piece to tether. That's one of the pieces on an IndyCar that if you bump into somebody and you don't really wreck, they can change quickly and get you back out there in the race. Not to say that should ever take the place of safety, but I'm sure that's why . . . there were probably a lot of teams or people who lobbied against getting them tethered because it makes it harder to switch that out in case of a small accident where you could keep on going.

I'm sure that if that was on people's radar of what they should do to make the sport safer, I'm sure there were people who thought that . . . Nobody thought that that was potentially going to harm somebody.

I think that there are times when it's easy for someone to find speed because of where the cars are, as far as the downforce level goes, but to be able to win races and win championships it still takes an "elite talent." You look at the guys who generally win races (win big races)—they're top-level drivers. But you've always had guys who are able to buy their way in to a race and maybe weren't the top level. When the cars are a little easier to drive it just allows them to get into a car that may be capable of running in the top five or lead the race and have an issue like that. I know that there are a lot of people who think [these cars] are easy to drive. There are probably some ways that it is easier, but to win races, it still takes good, elite talent. Generally, you don't see people winning any races where you say, "Man, that guy shouldn't have won"—except for maybe the "fuel mileage" thing—but you're never going to fix that.

## Q: TOP INDY MEMORY?

My favorite memory from Indy was probably the year that I won. That year we went into it . . . We were finally on a level playing field, as far as everybody having Hondas. That was what the "power house" was at that point. We went there, [and] we were the fastest every day of that month of May, except for one when we were second fastest.

The way that I was able to win the race, I was able to get through the last five laps without bawling my eyes out, thinking about trying to win the 500. Still, I was a bit surprised—or amazed—that we were able to get the job done. Having the "pit road issue"* that we did and to be able to come back from being almost a lap down and to restart, basically, ninth on the track for the last restart with four laps to go, and to be able to win the race, it was really magical.

I've had a lot of great memories there from inside the car and I've had some as a fan. I always cherish going back to the Speedway. Mostly, it reminds me of being a kid. Every time I drive through those gates, I generally don't think about going to Victory Lane. I think about walking there from over there by turn two where the big gas storage tanks are that Marathon has and walking, being a six-, ten-, twelve-, or thirteen-year-old kid. Walking there with my dad or with my parents. Just thinking about getting to spend the day with them, eating cold fried chicken, having a pop, and watching cars go by at 200 and something miles an hour. It's more about taking me back to being a kid than it ever is about driving a race car there. Because that's why I did it. I loved being part of . . . I wanted to be part of the greatest race in the world. I never would have thought I'd have the opportunity to win it. . . .

A lot of people ask me, "Why didn't you just stay in IndyCars? You might have been the first five-time winner. You could have been the greatest of all time. If your win percentage in IndyCars is as high up there as anybody has ever been . . . You could have . . ."

I know, but I might not have ever won another race. You don't know how it's going to happen. It took me . . . out of eight years that I ran the Speedway I only finished on the lead lap twice. It took me seven times to be able to win. I know I could have done it for another thirty-five years, and the odds are, you don't win it five times.

My kids have been there, but never to the Indy 500. My son is totally into "wheels." I've taken him to the museum there and let him check out cars and stuff like that. He's still a little bit too young for me not to be holding on to him . . . he'd be trying to climb into one.

We have my first IndyCar. That G-Force that I ran to do my rookie test. My dad bought that from the team that I drove for and had it restored to the way that it was. It sits in the lobby of my dad's business. My son, any time we go to see my mom while she's working, he is all over that thing. I'm trying to pull him off and they're like, "He can't hurt it, and if he does, we'll get it fixed." He loves it.

I know they'll be to an Indy 500. I don't know when. That time is gonna come.

In 2014 when I was running for "Gibbs" [Joe Gibbs Racing], we had an engine fail at Chicago on lap 7.

My oldest daughter loves going to [Chicagoland Speedway] because she gets to go up in the grandstands. One of our friends, he's like their uncle . . . he always takes her out to the grandstands to watch the race.

She came down [to the pit] for the start of the race. By the time they got back to their seats, the engine had blown up and she was like so bummed out. Dad was out of the race already. I thought, *This could be really bad, but* . . . I went up and sat in the grandstands with her. As mad and as bad as I felt a half-hour ago? I could just sit there and eat popcorn with her and watch the race with her.

She said: "Dad, this is like the best day ever."

*A "drive through" penalty assessed on lap 163 left Hornish thirty seconds behind the race leader.*

Sam pulls on the most conservative (and hottest) flame-resistant head sock.

Victory at California Speedway.

Sam with Speedway Engine's Jeff Gordon prepares for a practice run in the Panther Racing #4 car.

Panther Racing's transporter or "hauler."

Sam had a very successful run with Panther Racing, including a "points" championship. Here, in his final race with Panther, his crew has put a "Thanks Sammy" message behind the steering wheel. Sam would move on the Penske Racing, with whom he won the 500 in 2006. He passed Rookie Marco Andretti on the front straightaway in the 200th lap to win by a margin of less than a tenth of a second.

## 38
# EDDIE CHEEVER

*Won Indy 1998*

*Date of interview: October 2015*

**Q: PURSES AND PRIZE MONEY—HOW IS THAT DIVIDED UP?**

That changes from team to team and according to circumstances. Everybody *knows* how much the winner gets; just look in the paper the next day. Every team has a different way to go about it. When I won I was driving for myself, so there was never any question about how to divide the proceeds in the correct manner with the guys. There are certain managers that have to receive a larger share than certain mechanics. There are a variety of considerations. I don't think you can pinpoint one distribution pattern and say that's "common" across the whole racing platform.

**Q: THE 2015 RACE SAW JAMES HINCHCLIFFE GETTING INJURED AFTER QUALIFYING BUT BEFORE THE RACE. RYAN BRISCOE ULTIMATELY WAS CHOSEN TO FILL IN, DRIVING THE NO. 5 CAR. WHAT FINANCIAL CONSIDERATIONS WOULD BE INVOLVED IN A SITUATION LIKE THAT?**

I have absolutely no idea. To answer that question you'd have to be privy to the team's balance sheet and what their needs are at that time. Obviously the "later in the day" you change your driver at Indy, the numbers change, but they can swing either way. It's not as cut-and-dried as you're making it out to be. It's a very fluid situation—those things change. What's the driver's availability? What's the sponsorship package? And like any business, the team owners will do everything they can to insure a successful Indy 500, but there comes a point . . . you have to make sure you cover your "nut" accordingly, see that your "downside" is taken care of, because there's always the potential for . . . destroying the car. That hurts your balance sheet!

## Q: IS IT POSSIBLE TO INSURE A RACE CAR?

We often did. It's a very high premium, but yes. Don't forget there is a deductible just like anything. If you were to have a big accident, it's worthwhile. But . . . with the higher quality drivers . . . it's not an easy formula, either. Who's driving the car? How is the car prepared? What's the "history" with the car? There's a lot of numbers and information that's available to everyone, so . . . whoever's going to underwrite that obviously has a lot of knowledge. There's a company out of France that does it. We did it when we were racing, but we haven't owned race cars for a long time, so . . .

## Q: YOU WERE "OWNER/DRIVER" WHEN YOU WON IN 1998. DID YOU ALSO HAVE LOTS OF SPONSOR BACKING?

I didn't . . . I didn't.

I did a deal with Rachel's Potato Chips the second week of qualifying. We didn't have . . . it was a difficult month for us going into the Indy 500.

## Q: DRIVERS DON'T EVER TALK ABOUT HOW MUCH THEY'VE BEEN PAID (OR EARNED) DRIVING A RACE CAR.

There's probably a reason for that!

I'm sure . . . when you win in racing you're well paid.

When you *don't* win in racing, you're *less* well paid, like any other sport.

## Q: CAN YOU GUESSTIMATE THE COST OF CAMPAIGNING A ONE- OR TWO-CAR TEAM FOR INDY OR THE WHOLE IRL SEASON?

Again, I have no idea because I haven't been doing it for the last decade or so. I don't really have an idea.

I mean, you could figure out those numbers. Call Dallara and find out what that chassis costs. Get the information on the leasing of the engine (from Chevrolet or Honda) . . . top management? I'd say anywhere from $250,000 down to $110,000. Count all the people you're paying . . . the equity alone is probably two million dollars to get ready to run in that capacity. Putting together a team, assets, cars, trucks, all that stuff, it's not a complicated number to figure out.

## Q: DO DEEP POCKETS AND A READY INVENTORY OF SPARE PARTS GIVE GREAT ADVANTAGE OVER A SMALLER, LESS WELL-FINANCED TEAM?

We're talking about Indy? You actually have three goals at Indy. The first goal is to qualify . . . *to qualify*! Period!

The second goal, you want to qualify as *well* as you can, but it becomes irrelevant, trying . . . you're putting the equipment too much at risk if you're not, if you don't have a shot at the first two rows.

Once you have qualified, you have to start the race. You're not going to collect any prize money, and I *think* . . . the "last place" position earns over $200,000 now. So that goes a long way to making your "month" more profitable. The worst thing that would happen, such as the example of a team that doesn't have enough spare parts, you can't prepare the car . . . you don't have accidents at Indy where you just break the nose of the car. It's very rare that the accident is that small. You usually take away the gearbox or half the car, the chassis . . . it's always a very salty repair bill at Indy because the speeds are so high. So if you scrounge together a car at the last minute and enough spares to get yourself into the race, you've really achieved everything that you're going to achieve; the rest is luck.

That's one approach.

The other approach is the Penske approach where . . . you *have* to win! They spend enormous amounts of money making sure they have backup programs, making sure they have redundancy programs, one after another, that just fall into place when there's an issue . . . and in that case, it's a multimillion-dollar budget. Those are the items you have to cover.

### Q: PENSKE HAD TO DEAL WITH A DAMAGED PART OF THE BACK OF MONTOYA'S CAR IN THIS YEAR'S (2015) RACE . . .

That was nothing. That's like having a hiccup during a meal. He recovered and went on to win the race. The team was very well prepared. Whatever piece they put on the back [right rear impact attenuator or fender] was just . . . they had practiced that multiple times. Now a much smaller team would not have been able to change that piece as critically and efficiently as Montoya. They, Penske, probably practiced changing that piece a hundred times over the winter.

### Q: IN YOUR LONG AND VARIED CAREER AS A DRIVER, YOU ONCE DROVE FOR A. J. FOYT . . .

I got along very well with A. J. He's obviously got an enormous amount of experience from . . . I think it was four decades of racing, incredibly successful as a driver, a strong personality.

I was sorry we didn't win that race. We were leading with two laps to go at Nazareth [April 1995], but I enjoyed the experience

a lot. [He's] one of America's greatest race-car drivers. For me, plugging into that knowledge was a great experience. He's a rabid competitor. He desperately wants to win at everything he does—an argument, a discussion, a car—anything. That's just part of his DNA.

### GOLDEN INDY MOMENT:

I would say the best part of the race was those sixty seconds you have, when you've won the race and you're slowing down and coming into the pits. The variety of things that pass through your mind . . . in my case it was . . . my dad, all kinds of stuff, race cars. It's amazing how my mind was downloading all that stuff and more. I took my helmet off and had to say something intelligent to two hundred thousand people after I'd been dodging bullets for three and a half hours, but that was probably the most enjoyable part of the 500—that moment, right after, when you realize that you just won.

Eddie Cheever, Allen Bestwick, and Scott Goodyear are en route to the broadcast booth to cover the 500 for ABC/ESPN.

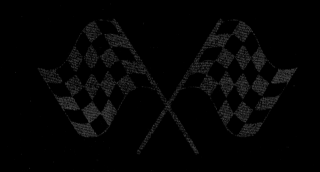

# 39
# JOHNNY RUTHERFORD

*Won Indy 1974, 1976, 1980*
*Date of interview: January 2004*

Rutherford's image for his 1976 victory on the Borg-Warner trophy.

**THE FIRST** time I addressed the starting field as a pace car driver, I told the guys, "Now look, there is absolutely nothing that you can think of back there that will 'bugger' a pace car or cheat the pace car in any way. Because I've already been back there and done all that. There is nothing that you can think of that I haven't already attempted. So, *don't mess with me* [*laughs*]." The fact that I raced for nearly forty years [means that] I have been back there; I know what it's like. I've seen the tricks pulled and the things done—trying to get extra position or laps or whatever with the pace. So I feel very comfortable in what I'm doing driving the pace car. It is an integral part of the event.

Sometimes you have a caution flag. They want to pack the field up and slow down, open the pits, and do the things they need to do when the caution comes out. That and the start. We've been working on trying to get good, clean, picture-perfect starts, which is nearly impossible. Someone will always pull out of line or try to win it in the first turn. Of course, Brian Barnhart always tries to impress that in the drivers meeting: "Guys, this is a 500-mile race; you can't win it in the first turn. There's no point in putting people in jeopardy. You can't be creating an unsafe condition."

Johnny Rutherford, known as "Lone Star JR," at the wheel of the Chevrolet Corvette pace.

It's something we try to work on. Invariably somebody steps out of line, but all in all the guys have been pretty good. It's been enjoyable for me, being "up close and personal" driving the pace car.

We have two-way radio communication between the pace car and "race control." We're controlled and given information from their vantage point, which is with all their monitors and everything. We're limited in our scope from where we "stage" with the pace car there at the first turn—at the end of the pits, normally. All we can see is the first turn, so we don't know what's happening at all the vantage points around the track. So they keep us up-to-date on that information. Our call is "Yellow-Yellow-Yellow" when the caution comes on. We know to stand by, look for the leader. They call the leader to us. We then pull out in front of the leader and slow the field down. For obvious

reasons—if there's an accident on the track, have the pace car slow the field down so that the safety workers can get to the incident to do whatever they have to do, either emergency-wise or just clean the situation up. If there's an oil spill, I'll obviously try to lead the field around that while they clean it up. Sometimes we'll go down on warm-up lane, the apron. It's not brain surgery or nuclear physics, it's just common sense. That's part of the duties.

I attend all of the drivers meetings so I can know what has transpired. Only in isolated incidents or special cases do I put in my two cents' worth, so to speak. At the 500 there is always a meeting after the drivers meeting of the "front row." That's always a very critical part of the start of the race—the three people there help setting the pace and bringing the field down for the green flag. Especially the pole position—the fastest qualifier is setting the pace coming down. I don't know what it is about drivers; they don't seem to know what the word "pace" means. Invariably, they want to lay back and play games with each other. They're playing these head games with each other and they don't keep up with the pace car. What is it about *pace* that you don't understand?

That's okay, we all did it. The pole position controls the pace, so he slows the field down so *he* can get a little bit of a "jump." It's a fun situation, the games they try to play. Occasionally they can come down and just do it right. There *are* some drivers that you can count on. I always have a meeting with the pole position driver just before he gets in his car, before he saddles up to start the race. I park the pace car in front of the field. I say to the "pole" driver, "That's the spacing we want—keep that spacing. You want to go faster? Crowd me, and I'll speed up. You want to go slower? Don't! I'm the pace car; you run my speed."

Particularly on a hot day, when you slow those things down they tend to heat up a little bit. If there's a "heating condition," we'll get it obviously from the crews along pit road, or from race control; they'll pass the word along. But the drivers will usually crowd me. You know, they'll get right up on me, want to "push" me or hand signal to "pick it up," or I'll up the pace to keep everything with the cars and drivers in "running order." When I bring the field down for the start I come off at the fourth turn at the start of the pit entrance. That gives the track to the cars and they come down to get the green flag. I go through the pits. I come through pit lane and usually down at the end there's an area at the end of the pits where we stage the pace car. I just sit, pull up, back in, and we have our vantage point there. We are usually arranged so that they can see the pace car from "control." They're at the top of the press box in the grandstand. We pull in there, sit (with the engine running), and watch what happens in front of us. At Indianapolis, the fastest I've ever had to "pace" is about a hundred miles an hour. The shorter tracks we'd be running about sixty-five to eighty depending on the circumstances.

The fact that I've been a driver of some success aids me obviously in the pace car. We have weather conditions, maybe rain or sprinkles on the racetrack. If I'm out there, I can make judgments about track conditions, whether we should restart or how it should go. They rely on my judgment sometimes for that. It's raining hard, that's it, we're through. But if it's sprinkles . . .

The Rutherfords' "his 'n' her" checkered flag rings.

One of three Indy 500 rings on the race-gnarled hand of Johnny Rutherford.

Invariably on a big racetrack (on any racetrack), you can have more rain on one area of the racetrack than on the other, so on the press box in the grandstands they're not getting much on their windows, but maybe down in turn two or even turn three way back on the backstretch it's sprinkling pretty hard. You've got to watch the color of the racetrack; if it starts changing color you'd better "red flag" and wait out the situation.

I was doing a test once at Indy. It was a fuel mileage test and we had to run a steady sixty miles an hour around the track until the thing ran out of fuel. We had a glass jug bolted to the right front fender; I could see it. We put in exactly a gallon of fuel and I was to go out and run it "dry." A rather ominous-looking storm came up in the third and fourth turn at the north end of the racetrack—a lot of dark clouds and everything. The other end had sunshine on it (this was in the fall). Before I could finish the run out on the fuel test, it was dry and sunshiny in turns one and two, and there was six inches of snow in turns three and four [*laughs*]. That's how different it can be, and how big the Speedway is.

## ON BEING A THREE-TIME INDY CHAMPION:

It's been my career—to have won the Indianapolis 500. That's one of the unfortunate things in this life, is that the things that means so much to you personally, individually, you can't explain to anybody else. It's just *your* situation.

If you win the Indianapolis 500 it means a great deal. It's not just the personal satisfaction; it's a great sense of accomplishment, probably the greatest sense of accomplishment (or one of a few) that I've had in my life that really means something. So to win the race—individually, it means a great deal, but it's not an individual effort, it's a team effort. You make friendships. You are involved with people who have the same ideals and virtual personalities you do. You're all trying to achieve the same thing. It's a great feeling. You make friendships, fast friendships. The teams and people that I've worked with are still strong and very good friends, both inside and outside racing. Camaraderie, kinship—these things you cherish. For the Indy 500 to be a focal point and cause of this "situation" is something special. There are friends that I've made and there are things that I've been able to do because I won the Indianapolis 500.

# RACING WITH GEORGE

THE LITTLE girl's birthday is in early June—just a few days after the big race. "Will you be back in time for my party?" she always asks.

One year, before leaving for the trip to Indy, her photographer dad secretly slipped her little Curious George® toy into a pocket of his camera bag. These photos attest to that grand adventure.

Curious George® on the famous "Yard of Bricks" start / finish line.

Curious George® in the "pole position" at the start / finish line.

Kenny Brack—winner of the 1999 Indy 500.

Race fan "Uncle Frank" with Panther Racing co-owner Mike Griffin.

IndyCar driver Ryan Briscoe.

2006 Indy Winner Sam Hornish Jr.

Tony George—(Anton Hulman George); his family owns the Speedway.

Curious George® visiting Tony George's old team—Vision Racing.

Indy and NASCAR racing sensation Danica Patrick. "I had this toy as a kid," she said.

David Letterman—co-owner Rahal-Letterman Racing.

Four-time Indianapolis 500 winner Al Unser Sr.

# APPENDIX B
# ONCE, AROUND THE TRACK

The Indy 500 has an intimate feeling about it, even though there are several hundred thousand other folks in attendance, as well. A "typical fan" defies description. Expect to meet every type from every corner of the world—even celebs!

One of the great drivers who never managed to win Indy—but should have—Lloyd Ruby, shown here after competing in the Heroes of Indy event at Texas Motor Speedway. Sadly, Ruby passed away in 2009 at the age of eighty-one.

NASCAR racing great Rusty Wallace.

Panther Racing's Jane Barnes.

A steadfast supporter in Dario Franchitti's pit, Ashley Judd.

Rapper/entrepreneur Chris Bridges—a.k.a. Ludacris.

This gentleman, Jim Nabors, is beloved at Indy. People who mainly knew him as the Gomer Pyle character on *The Andy Griffith Show* were surprised at his rich baritone voice. On thirty-six race mornings he belted out "Back Home Again in Indiana." He chose to let 2014 be his final performance.

Broadcast pit reporter Jamie Little.

Golfing legend Jack Nicklaus.

Mega-successful race team owner Chip Ganassi.

Academy Award-winning actor Jon Voight. For the uninformed—his daughter is Angelina Jolie.

Honda Motorsports' T. E. McHale.

Another great driver who was never blessed with an Indy win, Michael Andretti.

A photographer's dream strolling the midway—two *Playboy* models.

Rupert Boneham, a contestant on an early *Survivor* TV show.

Actor and bona fide racing guy James Garner. Sadly James Garner passed away in 2014 at age eighty-six.

Florence Henderson is a virtual "fixture" at Indy each Memorial Day.

Here's a great racer who's deserving of a future victory at Indy, Sebastien Bourdais.

*Real* race fans know what a skilled driver is Vitor Meira.

Legendary driver Jackie Stewart.

Actor Morgan Freeman in the pre-race parade of VIPs.

Panther Racing's John Barnes.

Veteran Indy racer and Indy-winning team manager Lee Kunzman.

Hard-charging competitor Paul Tracy.

It will surprise no one if/when this racer, Townsend Bell, gets to Victory Circle.

This is Dick Matlock, the most senior "yellow shirt" at the Indianapolis Motor Speedway. For over fifty years he's been supervising the friendly, efficient (and mostly part-time) Yellow Shirts. These workers drive trams, jitneys, and golf carts; they answer fans' questions, check credentials, run elevators, chauffeur VIPs and the infirm, and perform a thousand other tasks at the sprawling venue. Many yellow shirts are retired or take time off from their regular jobs to earn a little cash and demonstrate the friendly, firm, and fair Hoosier hospitality. Matlock's radio crackles constantly with the oddest requests and he calmly "gets it done."

This is Dick Matlock's comedy sidekick: Harry Bunn (not to be confused with the soccer star of the same name). If you're parked out near the 30th Street "media overflow lot" and carry about fifty pounds of photo gear and need a free lift to an appointment in Gasoline Alley, I hope your golf-cart driver looks like this.

Racing reporter Curt Cavin. The *Indianapolis Star* journalist is the ultimate motor sports insider.

The Indianapolis "fan base" for ALL sports is easy to spot.

Budding motor sports reporter/photographer Ryan Kent Jr.

Alex Tagliani at the 2015 Indy 500. If you're racing for A. J. Foyt in the #48 car (a tribute to Dan Gurney) it can be assumed that in A. J.'s parlance: You're a "charger."

Indianapolis Motor Speedway president Doug Boles.

The young race director is setting up his "starting grid" on the Main Street sidewalk of Speedway, Indiana.

"The Captain" Roger Penske stops in his tracks to honor your request for a picture (in the mad moments following Montoya's amazing finish—Team Penske's 16th victory!). You decide that the image "wants in"—even if the focus is a bit out.

If you're assigned to the shooting platform high above the main grandstand, it's advised that you "pack in" everything you'll need for the long race. It takes about ten minutes just to climb up there.

Race fans often own or rent a scanner which allows them to officially "eavesdrop" (listen only, of course) on the "message traffic" that crackles around the racetrack. Teams and drivers have assigned frequencies— some are private, some are public. For a small fee, concessionaires can quickly program a handheld scanner with the competitors of the day. Heated language between a driver and his spotter, or a driver and his race strategist, can often be heard. Teams are aware that competitors could be listening in, and may affect decoy or dummy messages to disguise actual pit strategy, fuel consumption, car behavior, and other telling conditions.

Mario Andretti mentioned that this "flag" atop the scoring pylon was one of his main reference points in "reading" the wind conditions. Now the flag has been upgraded to a proper wind sock.

The first young driver that we lost early in this book project, Tony Renna.

The beloved Justin Wilson at the 2015 Indy 500. A few months later he would succumb to a traumatic brain injury he suffered during a race at Pocono Raceway in Pennsylvania. His friend and fellow competitor Graham Rahal has rallied financial support from Justin's legion of friends around the world to benefit his young family.

If it was Race Week, you might have expected to see smoke coming from Fred Nation's Blackberry.

Derrick Walker is the former President of Operations and Competition, IndyCar. The Scotland native (with a good part of his career taking place in Australia) was first known for "nuttin' and boltin'" on race cars of every type. Whether chief mechanic or race team owner, success seems to follow everything he's touched throughout his forty-year career.

With permission from Nicole (Mrs. Ryan Briscoe), of course, this is their daughter Finley at the 2014 Indy 500.

It's always good to ask first rather than "poach" a photographer's "set up" shot. Here (left), the thumbs-up indicates that permission has been secured. This (above) is but a portion of the men and women who look after all the "do's and don'ts" of Indy racing. The actual 2014 "panorama" that is being captured extends considerably to the "East and West."

By stepping back a few paces we're able to see the corps that waits for the 2014 winner Ryan Hunter-Reay to complete the 2½-mile victory lap and then step out of the open convertible for the various celebrations around the Yard of Bricks.

The "balloon release" is always a visual surprise shortly before race time. Every year (that we were there) the wind (fortuitously) wafts from East to West and brings the show directly across the front straightaway.

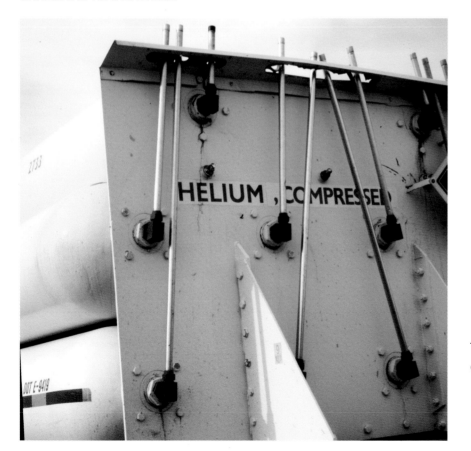

This "helium manifold" directs the balloon gas which has been stepped down in pressure to the various "filling stations" around the work tables.

There is obviously a pattern of filling clusters of the same color. As they are filled and "tied off" they are allowed to rise to the ceiling of the tarpaulin-covered shed. There is a party atmosphere that seems to bring the same workers back year after year. On cue, on race morning, the "roof" is slid away and off they go. The balloon building (bottom left) is a temporary structure usually located on the infield behind the "Pagoda" and Media Center.

Anthony Meadows USMC (left) was injured in action in the Middle East. Sam Schmidt (right) was a rising star in Indy racing with three successive starts in the 500. A racing accident in early 2000 as he was preparing for the coming season left him a quadriplegic. His Schmidt Peterson Motorsports is a front-line contender in Indy Racing.

ay be the favorite.
or boys are hot right now," for-
Noah said.

team remaining in the N.C.A.A.
as needed no last-second shots,

ntinued on Page D3

Elsa/Getty Images
Joakim Noah had 21 points, 15 rebounds and 5 blocked shots to lead third-seeded Florida past top-seeded Villanova.

Continued on Pag...

## I.R.L. Rookie Dies After Prerace Collision

**By DAVE CALDWELL and CHARLIE NOBLES**

The Indy Racing League rookie Paul Dana died yesterday after he was involved in a two-car collision during a practice session five hours before the season-opening race at Homestead-Miami Speedway in Florida.

Dana was pronounced dead at noon yesterday at Jackson Memorial Hospital in Miami. He would have turned 31 on April 15.

A native of St. Louis, Dana was the newest driver on the Rahal Letterman Racing team, which also includes the 2004 Indianapolis 500 champion Buddy Rice and last year's IndyCar Series rookie of the year, Danica Patrick. The Hall of Fame driver Bobby Rahal and the late-night talk-show host David Letterman own the team.

Rice and Patrick withdrew from yesterday's race, the Toyota Indy 300, which took place as scheduled.

"It is a very black day for us," Rahal said at a news conference before the race, which was won by Dan Wheldon, last year's I.R.L. champion.

Dana was fatally injured when his car slammed into the car driven by Ed Carpenter, the stepson of the I.R.L. founder, Tony George. Carpenter's car had gone into a spin on the second turn on the 1.5-mile oval. It then slid down the high banking and came to a stop near the bottom of the track before it was hit by Dana's car, which was traveling at nearly 200 miles an hour.

The collision occurred at 10:03 a.m. yesterday, two minutes into the final practice session before the race. Brian Barnhart, president and chief operating officer of the I.R.L., said the practice session was the first time in which all 20 cars were on the track at the same time.

Continued on Page D8

Luis Alvarez/Associated Press
Dana, 30, was a rookie in the Indy-Car Series.

This is the clip from the *New York Times* noting the death of Paul Dana.